SEMIHYPERGROUP THEORY

SEMIHYPERGROUP THEORY

BIJAN DAVVAZ
Department of Mathematics, Yazd University, Yazd, Iran

ELSEVIER

AMSTERDAM • BOSTON • HEIDELBERG • LONDON
NEW YORK • OXFORD • PARIS • SAN DIEGO
SAN FRANCISCO • SINGAPORE • SYDNEY • TOKYO
Academic Press is an imprint of Elsevier

Academic Press is an imprint of Elsevier
125 London Wall, London EC2Y 5AS, UK
525 B Street, Suite 1800, San Diego, CA 92101-4495, USA
50 Hampshire Street, 5th Floor, Cambridge, MA 02139, USA
The Boulevard, Langford Lane, Kidlington, Oxford OX5 1GB, UK

Notices

Knowledge and best practice in this field are constantly changing. As new research and experience broaden our understanding, changes in research methods, professional practices, or medical treatment may become necessary.

Practitioners and researchers must always rely on their own experience and knowledge in evaluating and using any information, methods, compounds, or experiments described herein. In using such information or methods they should be mindful of their own safety and the safety of others, including parties for whom they have a professional responsibility.

To the fullest extent of the law, neither the Publisher nor the authors, contributors, or editors, assume any liability for any injury and/or damage to persons or property as a matter of products liability, negligence or otherwise, or from any use or operation of any methods, products, instructions, or ideas contained in the material herein.

Library of Congress Cataloging-in-Publication Data
A catalog record for this book is available from the Library of Congress

British Library Cataloguing-in-Publication Data
A catalogue record for this book is available from the British Library

ISBN 978-0-12-809815-8 (print)
ISBN 978-0-12-809925-4 (online)

Publisher: Glyn Jones
Acquisition Editor: Glyn Jones
Editorial Project Manager: Tessa de Roo
Production Project Manager: Debasish Ghosh
Designer: Greg Harris

Typeset by SPi Global, India

CONTENTS

PREFACE

A semigroup is an algebraic structure that consists of a nonempty set together with an associative binary operation. The formal study of semigroups began in the early 20th century. Semigroups are important in many areas of mathematics because they are the abstract algebraic underpinning of "memoryless" systems: time-dependent systems that start from scratch at each iteration. In applied mathematics, semigroups are fundamental models for linear time-invariant systems. In partial differential equations, a semigroup is associated with any equation whose spatial evolution is independent of time. The theory of finite semigroups has been of particular importance in theoretical computer science since the 1950s because of the natural link between finite semigroups and finite automata. In probability theory, semigroups are associated with Markov processes.

Semihypergroups represent a natural extension of semigroups and they were introduced in 1934 by the French mathematician F. Marty. Since then, hundreds of papers and several books have been written on this topic. In a semigroup, the composition of two elements is an element, while in a semihypergroup, the composition of two elements is a nonempty set. Semihypergroups have many applications in automata, probability, geometry, lattices, binary relations, graphs, hypergraphs, and other branches of science such as biology, chemistry, and physics. The books published so far on hyperstructures deal especially with hypergroups, hyperrings, and polygroups. That is why I think that a book on semihypergroup theory is a necessity, especially for the new researchers on this topic and for Ph.D. students who need a material on semihypergroups. The idea to write this book, and more importantly the desire to do so, is a direct outgrowth of a course I gave in the department of mathematics at Yazd University. One of my main aims is to present an introduction to this recent progress in the theory of semihypergroups. I have tried to keep the preliminaries down to a bare minimum. The book covers most of the mathematical ideas and techniques required in the study of semihypergroups.

The presented book is composed of five chapters. In the first chapter, we recall some notions and basic results on semigroup theory that we shall extend to the context of semihypergroups. Chapter 2 is about semihypergroups, history of algebraic hyperstructures, and some basic results especially on some important classes of semihypergroups.

We study hyperideals, quasihyperideals, prime and semiprime hyperideals, semihypergroup homomorphism, regular and strongly regular relations, simple semihypergroups, and cyclic semihypergroups. In Chapter 3, the concept of ordered semihypergroups and some examples are presented. Then, hyperideals, prime hyperideals of the Cartesian product, right simple ordered semihypergroups, and ordered semigroups derived from ordered semihypergroups are studied. In Chapter 4, fundamental relations defining semihypergroups are studied. By using the notion of fundamental relations, we can connect semihypergroups to semigroups. More exactly, starting with a semihypergroup and using the fundamental relation, we can construct a semigroup structure on the quotient set. Finally, in the last chapter we summarize the contributions of this book.

Bijan Davvaz
Department of Mathematics, Yazd University, Yazd, Iran

CHAPTER 1

A Brief Excursion Into Semigroup Theory

1.1 BASIC DEFINITIONS AND EXAMPLES

We give here some basic definitions and very basic results concerning semigroups.

Let S be a non-empty set and $\zeta : S \times S \to S$ a *binary operation* that maps each ordered pair (x, y) of S to an element $\zeta(x, y)$ of S. The pair (S, ζ) (or just S, if there is no fear of confusion) is called a *groupoid*.

Definition 1.1.1. A *semigroup* is a pair $(S, \cdot)$ in which S is a non-empty set and $\cdot$ is a binary associative operation on S, ie, the equation

$$(x \cdot y) \cdot z = x \cdot (y \cdot z)$$

holds for all $x, y, z \in S$.

For an element $x \in S$, we let x^n be the product of x with itself n times. So, $x^1 = x$, $x^2 = x \cdot x$, and $x^{n+1} = x \cdot x^n$ for $n \geq 1$.

A semigroup S is *finite* if it has only a finitely many elements. A semigroup S is *commutative*, if it satisfies

$$x \cdot y = y \cdot x$$

for all $x, y \in S$. If there exists e in S such that for all $x \in S$,

$$e \cdot x = x \cdot e = x$$

we say that S is a *semigroup with identity* or (more usual) a *monoid*. The element e of S is called *identity*.

Proposition 1.1.2. *A semigroup can have at most one identity.*

Proof. If e and e' are both identities, then $e = e \cdot e' = e'$. □

By Proposition 1.1.2, the identity element is unique and we shall generally denote it by 1. The description of the binary operation in a semigroup $(S, \cdot)$ can be carried out in various ways. The most natural is simply to list all results of the operation for arbitrary pairs of elements. This

Semihypergroup Theory
http://dx.doi.org/10.1016/B978-0-12-809815-8.00001-2

method of describing the operation can be presented as a *multiplication table*, also called a *Cayley table*.

Example 1

(1) Let $S = \{a, b, c\}$ be a set of three elements and define the following table.

$\cdot$	a	b	c
a	a	b	c
b	b	a	c
c	c	b	c

Then, S is a finite semigroup.

(2) Let $\mathbb{N} = \{0, 1, \ldots\}$ be the set of all non-negative integers and $\mathbb{N}^* = \{1, 2, \ldots\}$ the set of all positive integers. Then, $(\mathbb{N}, \cdot)$ is a semigroup for the usual multiplication of integers. Also, $(\mathbb{N}, +)$ is a semigroup, when $+$ is the ordinary addition of integers. Define $(\mathbb{N}, \star)$ by $n \star m = \max\{n, m\}$. Then, $(\mathbb{N}, \star)$ is a semigroup, since

$$\begin{aligned} n \star (m \star k) &= \max\{n, \max\{m, k\}\} = \max\{n, m, k\} \\ &= \max\{\max\{n, m\}, k\} = (n \star m) \star k. \end{aligned}$$

(3) Let S be a non-empty set. There are two simple semigroup structures on S: with the multiplication given by $x \cdot y = x$, for all $x, y \in S$, in this case, the semigroup $(S, \cdot)$ is called the *left zero semigroup* over S. Also, fixing an element $a \in S$ and putting $x \cdot y = a$, for all $x, y \in S$, gives a semigroup structure on S.

(4) The set $E(A)$ of functions on a set A, with functional composition, is a semigroup.

(5) The set $M_n(\mathbb{R})$ of $n \times n$ square matrices over real numbers, with matrix multiplication, is a semigroup.

(6) Let A be any set and $\mathcal{P}_f(A)$ be the set of all finite non-empty subsets of A. Then, $\mathcal{P}_f(A)$ is a semigroup under the operation of taking the union of two sets.

(7) Let I, J be two non-empty sets and set $T = I \times J$ with the binary operation

$$(i, j) \cdot (k, l) = (i, l).$$

Then, $(T, \cdot)$ is a semigroup called the *rectangular band* on $I \times J$. The name derives from the observation that if the members of $I \times J$ are pictured in a rectangular grid in the obvious fashion, then the product of two elements lies at the intersection of the row of the first member and the column of the second.

(8) The *direct product* $S \times T$ of two semigroups $(S, \cdot)$ and $(T, \circ)$ is defined by

$$(s_1, t_1) \star (s_2, t_2) = (s_1 \cdot s_2, t_1 \circ t_2) \quad (s_1, s_2 \in S, t_1, t_2 \in T).$$

It is easy to show that the defined product is associative and hence the direct product is, indeed, a semigroup. The direct product is a convenient way of combining two semigroup operations. The new semigroup $S \times T$ inherits properties of both S and T.

(9) The *full transformation monoid* $\mathcal{T}_X$ on a set X. This is the monoid of all mappings of the set X to itself. The operation is composition of mappings. This is a very important semigroup because it is the semigroup analog of the symmetric group S_X. For example, recall that Cayley's theorem tells us that every group can be embedded in some symmetric group; there is an analogous theorem for semigroups which tells us that every semigroup can be embedded in some full transformation monoid.

(10) The *bicyclic semigroup* $B = \mathbb{N} \times \mathbb{N}$ with binary operation

$$(a, b) * (c, d) = (a - b + \max\{b, c\}, d - c + \max\{b, c\}).$$

This is a monoid with identity $(0, 0)$.

(11) On the Cartesian product $\mathbb{Z} \times \mathbb{Z}$, we define a binary operation as follows

$$(a, b) \star (c, d) = \begin{cases} (a - b + c, d) & \text{if} \quad b < c \\ (a, d) & \text{if} \quad b = c \\ (a, d + b - c) & \text{if} \quad b > c, \end{cases}$$

for all $a, b, c, d \in \mathbb{Z}$. The set $\mathbb{Z} \times \mathbb{Z}$ with such defined operation is a semigroup.

(12) In the set of all continuous functions of two variables x and y, we define in the square $0 \leq x \leq a$, $0 \leq y \leq a$, the following operation, which plays an important role in the theory of integrals. The result of this operation carried out for the functions $K_1(x, y)$ and $K_2(x, y)$ is the function

$$\int_0^a K_1(x, t) K_2(t, y)\, dt.$$

As follows easily from the simplest properties of integrals, this operation is associative. Thus, we obtain a semigroup.

(13) We consider the set of functions of one variable which are absolutely integrable for $0 \leq x < \infty$. In many branches of mathematics, one considers the operation in this set, the result of which for $f_1(x)$ and $f_2(x)$ is the function

$$\int_0^x f_1(t)f_2(x-t)dt.$$

One can show that this operation is associative and commutative.

(14) Let $(S, \cdot)$ be a semigroup, I, Λ be two non-empty sets, and P a matrix indexed by I and λ with entries $p_{i,\lambda}$ taken from S. Then, the *Rees matrix semigroup* $M(S, I, \Lambda, P)$ is the set $I \times S \times \Lambda$ together with the multiplication

$$(i, s, \lambda) \star (j, t, \mu) = (i, s \cdot p_{\lambda j} \cdot t, \mu).$$

Definition 1.1.3. Let $(S, \cdot)$ be a semigroup, which is not a monoid. Find a symbol 1 such that $1 \notin S$. Now, we extend the multiplication on S to $S \cup \{1\}$ by

$$\begin{aligned} &a \star b = a \cdot b \quad \text{if } a, b \in S, \\ &a \star 1 = a = 1 \star a \quad \text{for all } a \in S, \\ &1 \star 1 = 1. \end{aligned}$$

Then, $\star$ is associative. Thus, we have managed to extend multiplication in S to $S \cup \{1\}$. For an arbitrary semigroup S, the monoid S^1 is defined by

$$S^1 = \begin{cases} S & \text{if } \; S \text{ is a monoid,} \\ S \cup \{1\} & \text{if } \; S \text{ is not a monoid.} \end{cases}$$

Therefore, S^1 is "S with a 1 *adjoined*" if necessary.

1.2 DIVISIBILITY OF ELEMENTS

If the semigroup $(S, \cdot)$ is not a group, its operation is not invertible. This means that for some elements, $a, b \in S$, there are no elements x, y such that

$$x \cdot b = a \quad \text{and} \quad b \cdot y = a.$$

It is possible, of course, that for some pairs of elements a and b one or other of those equations will have a solution with x or y in S. The question of which pairs of elements admit a solution, and which do not, is of the utmost importance in the study of the structure of semigroups and in the investigation of the properties.

Definition 1.2.1. Let $(S, \cdot)$ be a semigroup. An element b of the semigroup S is called a *right divisor* of the element a of the same semigroup if there exists an element $x \in S$ such that

$$x \cdot b = a.$$

b is called a *left divisor* of a if there exists an element $y \in S$ such that

$$b \cdot y = a.$$

If b is a right divisor of a, we say that a is *divisible on the right* by b. If b is a left divisor of a, we say that a is *divisible on the left* by b.

In the following, we present elementary properties of the relationship of divisibility in semigroups.

(1) If b is a right divisor of a, and c is a right divisor of b, then c is a right divisor of a.
(2) The product $a_1 \cdot a_2$ is divisible by a_1 on the left and by a_2 on the right.
(3) If b is a right divisor of a, then for arbitrary $z \in S$, the element $z \cdot a$ is divisible on the right by b.
(4) If a is divided on the right by b and b is divisible on the right by a, then b is a right divisor of itself.
(5) In order that the element b should be a right divisor of the element a, it is necessary and sufficient that in the multiplication table of S the column corresponding to the element b contains the element a.
(6) In order that the element b should be a left divisor of the element a, it is necessary and sufficient that in the multiplication table of S the row corresponding to the element b contains the element a.
(7) An element a of a semigroup S will be a right divisor of every element of S if and only if in the column of the multiplication table corresponding to the element a, all elements of S occur.
(8) a will be a left divisor of every element of S if and only if in the row of the multiplication table corresponding to a, all elements of S occur.
(9) The element a is divisible on the right by every element of S if and only if a occurs in every column of the multiplication table.
(10) a is divisible on the left by every element of S if and only if a occurs in every row of the multiplication table.

In the study of the property of divisibility of elements, there arises a question which is of interest under some circumstances: what are the properties of the element that multiplies the divisor to yield the dividend?

Definition 1.2.2. Let $(S, \cdot)$ be a semigroup. If the element b of S is a right divisor of the element a of the same semigroup, then the element x satisfying the equation

$$x \cdot b = a$$

is called a *left inverse* of b with respect to a.

The notion of *right inverse* is defined analogously.

An element which is both a right inverse and a left inverse of b with respect to a is called a *two-sided inverse*, or shortly an *inverse*, of b with respect to a.

The properties listed below follow from the definitions.

(1) If the element x is a left inverse of b with respect to a, then for arbitrary z belonging to the given semigroup, x is a left inverse of $b \cdot z$ with respect to $a \cdot z$.

(2) If the element x is a left inverse of b with respect to a, then for an arbitrary element z of the given semigroup, $z \cdot x$ is a left inverse of b with respect to $z \cdot a$.

(3) If x_1 is a left inverse of b with respect to x_2, and x_2 is a left inverse of c with respect to a, then x_1 is a left inverse of $b \cdot c$ with respect to a.

(4) The number of the left inverses of an element b with respect to a is equal to the number of times that a appears in the column of the multiplication table of the semigroup corresponding to b.

(5) The number of the right inverses of an element b with respect to a is equal to the number of times that a appears in the row of the multiplication table of the semigroup corresponding to b.

The following particular case is of special interest.

Definition 1.2.3. In a semigroup $(S, \cdot)$ an element z_l such that $z_l \cdot x = z_l$ for every $x \in S$ is called a *left zero element* of S, and an element z_r such that $x \cdot z_r = z_r$ for every $x \in S$ is called a *right zero element* of S. If z is both a left and a right zero element of S, then z is called a *two-sided zero element*, or simply a *zero element*, of S.

A semigroup S may have any number of left (or of right) zero elements, but if it has a left zero element z_l and a right zero element z_r, then $z_l = z_l \cdot z_r = z_r$, whence S has a unique two-sided zero element and no other left or right zero element.

Definition 1.2.4. A semigroup $(S, \cdot)$ is *left cancellative*, if

$$c \cdot a = c \cdot b \Rightarrow a = b,$$

and $(S, \cdot)$ is *right cancellative*, if

$$a \cdot c = b \cdot c \Rightarrow a = b.$$

If $(S, \cdot)$ is both left and right cancellative, then it is *cancellative*.

Example 2

(1) The set of positive integers under addition is a cancellative semigroup.

(2) Consider the semigroup $M_2(\mathbb{R})$ of 2×2 square matrices over real numbers with matrix multiplication. Then,

$$\begin{pmatrix} 1 & 0 \\ 0 & 0 \end{pmatrix} \cdot \begin{pmatrix} 1 & 1 \\ 1 & 1 \end{pmatrix} = \begin{pmatrix} 1 & 0 \\ 0 & 0 \end{pmatrix} \cdot \begin{pmatrix} 1 & 1 \\ 0 & 0 \end{pmatrix},$$

and so $M_2(\mathbb{R})$ is not a cancellative semigroup. However, if we take all matrices A with $det(A) = 1$, then this semigroup is cancellative.

1.3 REGULAR AND INVERSE SEMIGROUPS

Regular semigroups were introduced by Green in his influential paper [1]. Also, see [2–4]. The concept of regularity in a semigroup was adapted from an analogous condition for rings, already considered by J. von Neumann. It is useful in a number of problems in the theory of semigroups.

Definition 1.3.1. Let $(S, \cdot)$ be a semigroup. We say that S is *regular* if for every $a \in S$, there exists $x \in S$ such that

$$a \cdot x \cdot a = a. \tag{1.1}$$

Example 3

(1) A rectangular band $I \times J$ is regular, since $(i,j) = (i,j) \star (k,l) \star (i,j)$, for all $(i,j), (k,l) \in I \times J$.

(2) The full transformation semigroup $\mathcal{T}_X$ is regular.

(3) The semigroup $(\mathbb{N}, +)$ is not regular, since there is no solution (in $\mathbb{N}$) to $1 + x + 1 = 1$.

Proposition 1.3.2. *A semigroup $(S, \cdot)$ is regular if and only if for every $a \in S$, there exists a' such that*

$$a \cdot a' \cdot a = a \quad and \quad a' \cdot a \cdot a' = a'. \tag{1.2}$$

Proof. Suppose that S is regular. Simply take $a' = x \cdot a \cdot x$. Then, from Eq. (1.1), we have $a \cdot a' \cdot a = a$ and $a' \cdot a \cdot a' = a'$.

The converse is clear. □

The element a' is usually called a *generalized inverse* of a.

Given a semigroup $(S, \cdot)$, we define the subset $V(a)$ of S by

$$V(a) = \{a' | a' \text{ is a generalized inverse for } a\}.$$

In a regular semigroup $(S, \cdot)$, every element has at least one generalized inverse. To see this, let us take an element $a \in S$. By definition, there is

$x \in S$ such that $a = a \cdot x \cdot a$. In this case, it is not only x, but also $x \cdot a \cdot x$ is a generalized inverse for a, since we have

$$a = a \cdot x \cdot a = a \cdot x \cdot (a \cdot x \cdot a) = a \cdot (x \cdot a \cdot x) \cdot a,$$
$$x \cdot a \cdot x = x \cdot (a \cdot x \cdot a) \cdot x = x \cdot a \cdot x \cdot (a \cdot x \cdot a) \cdot x = (x \cdot a \cdot x) \cdot a \cdot (x \cdot a \cdot x).$$

Example 4. Consider the semigroup $S = \{a, b, c, d\}$ with the following multiplication table.

$\cdot$	a	b	c	d
a	a	b	a	b
a	a	b	a	b
c	c	d	c	d
d	c	d	c	d

It is easy to check that every element is a generalized inverse of every other element.

Proposition 1.3.3. *Let $(S, \cdot)$ be a finite semigroup. Then, there exists $a \in S$ such that $a^2 = a$.*

Proof. Suppose that x is an arbitrary element of S. Since S is finite, it follows that $x, x^2, x^3, \ldots$ are distinct. So, there exist integers m, n with $n < m$ such that $x^m = x^n$. Hence, $x^{n+k} = x^n$, where $k = m - n$. Now, we have

$$x^{2n+k} = x^n \cdot x^{n+k} = x^{2n}.$$

By mathematical induction, we obtain

$$x^{rn+k} = x^{rn}, \quad \text{for all } r \in \mathbb{N}.$$

Also, we have

$$x^{rn+2k} = x^{rn+k} \cdot x^k = x^{rn} \cdot x^k = x^{rn+k} = x^{rn},$$

$$x^{rn+3k} = x^{rn+2k} \cdot x^k = x^{rn} \cdot x^k = x^{rn+k} = x^{rn},$$

and so on. Again, by mathematical induction, we obtain

$$x^{rn+lk} = x^{rn}, \quad \text{for all } l \in \mathbb{N}.$$

In particular, we obtain $x^{kn+nk} = x^{kn}$ or $x^{2nk} = x^{kn}$. Now, we set $x^{nk} = a$. □

The element $i \in S$ is called *idempotent* if $i^2 = i$. From Eq. (1.1), it is easy to see that both $a \cdot x$ and $x \cdot a$ are idempotent. Also, we define the set of idempotents in S to be

$$I(S) = \{i \in S | i^2 = i\}.$$

Note that $I(S)$ may be empty. In general, a semigroup consisting entirely of idempotents, ie, $I(S) = S$, is known as a *band*.

Example 5. Consider Example 1(7). Then, $(i,j)^2 = (i,j) \cdot (i,j) = (i,j)$, ie, every element is an idempotent.

Theorem 1.3.4. *The following conditions on a regular semigroup S are equivalent.*

(1) *Idempotents commute;*

(2) *Generalized inverses are unique.*

Proof. (1 ⇒ 2): Suppose that idempotents commute. Let a' and a'' be generalized inverses of a. Then, we obtain

$$\begin{aligned} a' &= a' \cdot a \cdot a' \\ &= a' \cdot a \cdot a'' \cdot a \cdot a' \\ &= a' \cdot a \cdot a'' \cdot a \cdot a'' \cdot a \cdot a' \quad \text{(by Eq. 1.2)} \\ &= a'' \cdot a \cdot a' \cdot a \cdot a'' \cdot a \cdot a' \\ &= a'' \cdot a \cdot a' \cdot a \cdot a' \cdot a \cdot a'' \quad \text{(by commuting idempotents)} \\ &= a'' \cdot a \cdot a'' \\ &= a''. \end{aligned}$$

(2 ⇒ 1): Suppose that i, i' are idempotents and x is the unique generalized inverse of $i \cdot i'$. We have

$$i \cdot i' \cdot x \cdot i \cdot i' = i \cdot i' \quad \text{and} \quad x \cdot i \cdot i' \cdot x = x.$$

Then, $i' \cdot x \cdot i$ is idempotent. Since

$$(i' \cdot x \cdot i)^2 = i' \cdot (x \cdot i \cdot i' \cdot x) \cdot i = i' \cdot x \cdot i,$$

and $i \cdot i'$ is a generalized inverse of $i' \cdot x \cdot i$, it follows that

$$\begin{aligned} &(i' \cdot x \cdot i) \cdot (i \cdot i') \cdot (i' \cdot x \cdot i) = i'(x \cdot i \cdot i' \cdot x) \cdot i = i' \cdot x \cdot i, \\ &(i \cdot i') \cdot (i' \cdot x \cdot i) \cdot (i \cdot i') = i \cdot i' \cdot x \cdot i \cdot i' = i \cdot i'. \end{aligned}$$

But an idempotent is its own unique generalized inverse and so $i \cdot i' = i' \cdot x \cdot i$, an idempotent. Similarly, $i' \cdot i$ is idempotent. The unique generalized inverse of $i \cdot i'$ is thus $i \cdot i'$ itself. On the other hand, $i' \cdot i$ is a generalized inverse of $i \cdot i'$, since

$$(i \cdot i') \cdot (i' \cdot i) \cdot (i \cdot i') = (i \cdot i')^2 = i \cdot i' \quad \text{and} \quad (i' \cdot i) \cdot (i \cdot i') \cdot (i' \cdot i) = (i' \cdot i)^2 = i' \cdot i.$$

It follows that $i \cdot i' = i' \cdot i$, as required. □

In connection with the fact that in a regular semigroup any element may have several generalized inverse, it is natural to consider separately the class of regular semigroups in which each element has a unique generalized inverse. Semigroups of this class have been the object of several studies since then; V. V. Vagner calls these semigroups "generalized groups." However, we prefer Preston's term, "inverse semigroups," as this reflects the most important properties of these semigroups.

Definition 1.3.5. A semigroup $(S, \cdot)$ in which every element a has precisely one generalized inverse is called an *inverse semigroup*.

To put this another way, $|V(a)| = 1$ for inverse semigroups.

Example 6

(1) The non-zero integers with addition form an inverse semigroup.
(2) The bicyclic semigroup is an inverse semigroup.
(3) A rectangular band $I \times J$ is an inverse semigroup only if it is trivial, ie, $|I| = |J| = 1$.
(4) The full transformation semigroup $\mathcal{T}_X$ is not an inverse semigroup.

In the following lemma, there are two interesting properties of inverse semigroups.

Lemma 1.3.6

(1) *Every inverse semigroup is regular.*
(2) *In every inverse semigroup, idempotents commute.*

Proof. It follows from Theorem 1.3.4 and Definition 1.3.5. □

1.4 SUBSEMIGROUPS, IDEALS, BI-IDEALS, AND QUASI-IDEALS

This section surveys properties of semigroups related to subsemigroups, ideals, bi-ideals, and quasi-ideals. A number of interesting examples are presented. We refer the readers to [5–15] for more details.

Definition 1.4.1. Let $(S, \cdot)$ be a semigroup and K a non-empty subset of S. Then, K is a *subsemigroup* of S if $a, b \in K$ implies $a \cdot b \in K$. If $(S, \cdot)$ is a monoid, then K is a *submonoid* of S if K is a subsemigroup and $1 \in K$.

A subsemigroup uses the operation of its mother semigroup, and hence if K is a subsemigroup of S, then certainly the operation of K is associative, and thus K is a semigroup by its own right.

Example 7

(1) A semigroup S is always a subsemigroup of itself.
(2) Consider the additive semigroup $(\mathbb{Q}, +)$ of rationales. Then, $(\mathbb{N}, +)$ is a subsemigroup of $(\mathbb{Q}, +)$.

(3) The subset with the property that its every element commutes with any other element of the semigroup is called the *center of the semigroup*. The center of a semigroup is actually a subsemigroup.

Lemma 1.4.2. *Let* $(S, \cdot)$ *be a semigroup and* $x \cdot y = y \cdot x$, *for all* x, y *in* $I(S)$, *the set of all idempotent elements of* S. *Then,* $I(S)$ *is empty or is a subsemigroup of* S.

Proof. Suppose that $x, y \in I(S)$. Then, we have $(x \cdot y)^2 = x \cdot y \cdot x \cdot y = x \cdot x \cdot y \cdot y = x \cdot y$. This implies that $x \cdot y \in I(S)$. □

Lemma 1.4.3. *Let* $\{K_i\}_{i \in \Lambda}$ *be a family of subsemigroups of a semigroup* S. *If their intersection is non-empty, then* $\bigcap_{i \in \Lambda} K_i$ *is a subsemigroup of* S.

Proof. It is straightforward. □

Remark 1. The union of subsemigroups need not be a subsemigroup. For example, in the additive semigroup of integers modulo 6, if K_1 consists of multiples of 2 and K_2 consists of multiples of 3, then $2 + 3 = 5 \notin K_1 \cup K_2$.

For a non-empty subset A of S, denote

$$B = \langle A \rangle = \bigcap \{K | K \text{ is a subsemigroup of } S \text{ such that } A \subseteq K\}.$$

By Lemma 1.4.3, $B = \langle A \rangle$ is a subsemigroup of S, called the *subsemigroup generated by* A, and A is called the *generating set* of B. It is the smallest subsemigroup of S that contains A. When A is a singleton set, $A = \{a\}$, then we write $\langle a \rangle$. More generally, if $A = \{a_1, \ldots, a_n\}$, then we write $\langle a_1, \ldots, a_n \rangle$.

Proposition 1.4.4. *Let* A *be a non-empty subset of a semigroup* S. *Then,*

$$\langle A \rangle = \bigcup_{n=1}^{\infty} A^n = \{a_1 \cdot \cdots \cdot a_n | n \geq 1,\ a_i \in A\}.$$

Proof. It is easy to see that $\bigcup_{n=1}^{\infty} A^n$ is a subsemigroup of S. Since $\langle A \rangle$ is a subsemigroup of S, it follows that A^n is a subset of $\langle A \rangle$, for all $n \geq 1$. So, $\bigcup_{n=1}^{\infty} A^n \subseteq \langle A \rangle$. Thus, the claim follows. □

There is an especially important case, namely when the subset A is a generating set for the semigroup S itself, ie, its generates the entire semigroups, or $S = \langle A \rangle$. Every semigroup, of course, has one or more generator sets.

We should note also the following self-evident properties:

(1) $\langle \langle A \rangle \rangle = \langle A \rangle$;

(2) If $A_1 \subseteq A_2$, then $\langle A_1 \rangle \subseteq \langle A_2 \rangle$.

Definition 1.4.5. If $A = \{a\}$ is a singleton subset of a semigroup S, then

$$\langle a \rangle = \{a,\ a^2,\ a^3, \ldots\}$$

is called a *cyclic subsemigroup* of S.

If $S = \langle a \rangle$ for some $a \in S$, then we say S is a *cyclic semigroup*.

Example 8. Let $M_2(\mathbb{Z})$ be the semigroup of 2×2 square matrices over integers. Suppose that

$$\alpha = \begin{pmatrix} 0 & 1 \\ -1 & 0 \end{pmatrix}.$$

Then, we obtain

$$\alpha^2 = \begin{pmatrix} -1 & 0 \\ 0 & -1 \end{pmatrix},\ \alpha^3 = \begin{pmatrix} 0 & -1 \\ 1 & 0 \end{pmatrix},\ \alpha^4 = \begin{pmatrix} 1 & 0 \\ 0 & 1 \end{pmatrix},\ \alpha^5 = \alpha.$$

Hence,

$$\langle \alpha \rangle = \{I,\ \alpha,\ \alpha^2,\ \alpha^3,\ \alpha^4\}$$

is a finite cyclic subsemigroup of $M_2(\mathbb{Z})$.

The *product* $A \cdot B$ of non-empty subsets A and B of a semigroup S is defined to be the set of all elements $a \cdot b$, where $a \in A$ and $b \in B$. It is almost an immediate consequence of the associativity of the multiplication of elements defined in S that the multiplication of subsets is associative, ie, $A \cdot (B \cdot C) = (A \cdot B) \cdot C$.

Definition 1.4.6. Let $(S, \cdot)$ be a semigroup and I be a non-empty subset of S.

(1) I is said to be a *left ideal* of S if $S \cdot I \subseteq I$;
(2) I is said to be a *right ideal* of S if $I \cdot S \subseteq I$;
(3) I is said to be a *two-sided ideal*, or simply an *ideal*, if it is simultaneously a left and a right ideal.

Example 9

(1) If a semigroup S has a zero element 0, then $\{0\}$ is an ideal of S.
(2) Let $\mathbb{Z}_{14} = \{0,\ 1,\ 2, \ldots, 13\}$ be the semigroup under multiplication modulo 14. Then, $I = \{0,\ 7\}$ and $J = \{0,\ 2,\ 4,\ 6,\ 8,\ 10,\ 12\}$ are ideals of $\mathbb{Z}_{14}$.
(3) If $i \in I$, then $\{i\} \times J$ is a right ideal in a rectangular band $I \times J$.
(4) Let $(S, \cdot)$ be a semigroup. For any $n \in \mathbb{N}$, the set

$$S^n = \{a_1 \cdot \cdots \cdot a_n | a_i \in S\}$$

is an ideal of S.

(5) Consider the semigroup consisting of the four elements

$$S = \{a,\ b,\ c,\ o\},$$

in which the product of any two elements is equal to o, with the exception of the one product $a \cdot b = c$. The associativity of the operation is verifiable without difficulty inasmuch as, for any three elements $x, y, z \in S$,

$$(x \cdot y) \cdot z = o \quad \text{and} \quad x \cdot (y \cdot z) = o.$$

The set $I = \{b,\ c,\ o\}$ is clearly an ideal of S. Its subset $J = \{b,\ o\}$ is an ideal of I. However, J is not even a left ideal of S.

In the following, we give some of the simplest properties of ideals.

(1) The union of any collection of ideals of a semigroup S is itself an ideal of S.

(2) The product of two ideals of S is an ideal of S.

(3) The intersection of any collection of ideals of a semigroup S is an ideal of S, if it is non-empty.

(4) The intersection of two ideals of S is an ideal of S.
Indeed, if I_1 and I_2 are two ideals of S, then their intersection is non-empty, since it clearly contains their product $I_1 \cdot I_2$. Then, according to (3), this intersection is an ideal of S.

(5) If S has a zero element 0, then 0 is included in every ideal of S.

(6) If K is a subsemigroup of S and I is an ideal of S, then $K \cap I$, if it is non-empty, is an ideal of the semigroup K.

(7) If I and J are left (right) ideals of a semigroup S with $I \cap J \neq \emptyset$, then $I \cap J$ is a left (right) ideal.

Definition 1.4.7. Let $(S, \cdot)$ be a semigroup.

(1) S is *simple* if it has no ideals $I \neq S$.

(2) If S has a zero element 0, then S is 0-*simple* if S and $\{0\}$ are the only ideals of S and $S^2 \neq 0$.

Example 10. The bicyclic semigroup B, defined in Example 1(10), is simple. Let I be an ideal of B and $(m, n) \in I$. Then, $(0, n) = (0, m) \star (m, n) \in I$. Thus,

$$(0, 0) = (0, n) \star (n, 0) \in I.$$

Now, suppose that $(a, b) \in B$. Then, we have

$$(a, b) = (a, b) \star (0, 0) \in I,$$

and so $B = I$.

Proposition 1.4.8. *Let $(S, \cdot)$ be a semigroup. Then, S is simple if and only if $S = S \cdot x \cdot S$, for all $x \in S$.*

Proof. Clearly, for any $x \in S$, $S \cdot x \cdot S$ is an ideal of S and so if S is simple, then $S = S \cdot x \cdot S$, for all $x \in S$.

Conversely, suppose that $S = S \cdot x \cdot S$, for all $x \in S$. Let I be an ideal of S and let $x \in I$. Then, $S = S \cdot x \cdot S \subseteq I$. So, $I = S$, from which it follows that S is simple. □

Remark 2. By Proposition 1.4.8, a semigroup S is simple if and only if for all $s, r \in S$ the equation $x \cdot s \cdot y = r$ has a solution in S.

Example 11. By using Remark 2, one can show that the semigroup of all 2×2 matrices,

$$\begin{pmatrix} x & 0 \\ y & 1 \end{pmatrix} \quad (x, y \in \mathbb{Q} \text{ with } x, y > 0),$$

is a simple semigroup.

Definition 1.4.9. The *principal left ideal* generated by an element $a \in S$ is the ideal $S \cdot a$. If S contains a left identity element e, then $a = e \cdot a \in S \cdot a$, but in the absence of a left identity element we may have $a \notin S \cdot a$ (for example, the principal ideal generated by 2 in the multiplicative semigroup of even integers). Dually, the *principal right ideal* generated by a is defined to be $a \cdot S$. If S contains a right identity element, then $a \in a \cdot S$. We also define the *principal ideal* generated by a to be $S \cdot a \cdot S$, which contains a if S contains a two-sided identity element.

Example 12. In the semigroup of positive integers under addition, every ideal is principal and has the form $I_n = \{n, n+1, n+2, \ldots\}$, for all $n \in \mathbb{N}$.

We make note of how the S^1 notation can be used. For example,

$$\begin{aligned} S^1 \cdot A &= \{s \cdot a | s \in S^1, a \in A\} \\ &= \{s \cdot a | s \in S \cup \{1\}, a \in A\} \\ &= \{s \cdot a | s \in S, a \in A\} \cup \{1 \cdot a | a \in A\} \\ &= S \cdot A \cup A. \end{aligned}$$

In particular, if $A = \{a\}$, then $S^1 \cdot a = S \cdot a \cup \{a\}$. Duality, $a \cdot S^1 = a \cdot S \cup \{a\}$ and similarly, $S^1 \cdot a \cdot S^1 = S \cdot a \cdot S \cup a \cdot S \cup S \cdot a \cup \{a\}$.

Lemma 1.4.10 (Principal Left Ideal Lemma). *Let $(S, \cdot)$ be a semigroup. The following statement are equivalent.*

(1) $S^1 \cdot a \subseteq S^1 \cdot b$;

(2) $a \in S^1 \cdot b$;

(3) $a = t \cdot b$ *for some* $t \in S^1$;
(4) $a = b$ *or* $a = t \cdot b$ *for some* $t \in S$.

Proof. It is straightforward. □

Lemma 1.4.11 (Principal Right Ideal Lemma). *Let* $(S, \cdot)$ *be a semigroup. The following statements are equivalent.*
(1) $a \cdot S^1 \subseteq b \cdot S^1$;
(2) $a \in b \cdot S^1$;
(3) $a = b \cdot t$ *for some* $t \in S^1$;
(4) $a = b$ *or* $a = b \cdot t$ *for some* $t \in S$.

Proof. It is straightforward. □

If A is an ideal of S, then $S \cdot A \cdot S = (S \cdot A) \cdot S \subseteq A \cdot S \subseteq A$. Conversely, if $S \cdot A \cdot S \subseteq A$ and S has a left identity element e, then $A \cdot S = e \cdot A \cdot S \subseteq S \cdot A \cdot S \subseteq A$, so that A is a right ideal of S; dually, if S has a right identity element and $S \cdot A \cdot S \subseteq A$, then A is a left ideal of S. Hence, if S has an identity element and $S \cdot A \cdot S \subseteq A$, then A is an ideal of S. But, in the absence of appropriate identity elements, $S \cdot A \cdot S \subseteq A$ need not imply either $S \cdot A \subseteq A$ or $A \cdot S \subseteq A$.

Example 13. Consider commutative semigroup $S = \{x_1, x_2, x_3, x_4, x_5, x_6\}$ whose multiplication table is exhibited below.

$\cdot$	x_1	x_2	x_3	x_4	x_5	x_6
x_1	x_1	x_1	x_1	x_4	x_4	x_4
x_2	x_1	x_1	x_1	x_4	x_4	x_4
x_3	x_1	x_1	x_2	x_4	x_4	x_5
x_4	x_4	x_4	x_4	x_1	x_1	x_1
x_5	x_4	x_4	x_4	x_1	x_1	x_1
x_6	x_4	x_4	x_5	x_1	x_1	x_2

Suppose that A is the subset $\{x_1, x_2, x_3, x_4\}$. Then, we have

$$S \cdot A \cdot S = S^2 \cdot A = \{x_1, x_2, x_4, x_5\} \cdot \{x_1, x_2, x_3, x_4\} = \{x_1, x_4\} \subseteq A,$$

but $A \cdot S = S \cdot A = \{x_1, x_2, x_4, x_5\} \nsubseteq A$.

However, it should be noticed that, for an arbitrary non-empty subset A of S, the subsets $S \cdot A$, $A \cdot S$, and $S \cdot A \cdot S$ are respectively left ideal, right ideal, and ideal of S.

Definition 1.4.12. Let $(S, \cdot)$ be a semigroup and B be a non-empty subset of S. We shall say that B is a *bi-ideal* of S if the following conditions hold.
(1) B is a subsemigroup of S;
(2) $B \cdot S \cdot B \subseteq B$.

The notion of quasi-ideal is a generalization of the notion of one-sided ideal. It was introduced by Steinfeld in 1953 for rings [16] and in 1956 for semigroups [17], and it has been widely studied in different algebraic structures. A monograph on quasi-ideals was written by Steinfeld in 1978 [18].

Definition 1.4.13. Let $(S, \cdot)$ be a semigroup and Q be a non-empty subset of S. We shall say that Q is a *quasi-ideal* of S if the following conditions hold.

(1) Q is a subsemigroup of S;

(2) $S \cdot Q \cap Q \cdot S \subseteq Q$.

Example 14

(1) Let

$$S = \left\{ \begin{pmatrix} 0 & 0 \\ 0 & 0 \end{pmatrix}, \begin{pmatrix} x & 0 \\ 0 & 0 \end{pmatrix}, \begin{pmatrix} 0 & y \\ 0 & 0 \end{pmatrix}, \begin{pmatrix} 0 & 0 \\ z & 0 \end{pmatrix}, \begin{pmatrix} 0 & 0 \\ 0 & t \end{pmatrix} \right\}$$

be a semigroup under multiplication of matrices. Then,

$$B_1 = \left\{ \begin{pmatrix} 0 & 0 \\ 0 & 0 \end{pmatrix}, \begin{pmatrix} x & 0 \\ 0 & 0 \end{pmatrix} \right\} \text{ and } B_2 = \left\{ \begin{pmatrix} 0 & 0 \\ 0 & 0 \end{pmatrix}, \begin{pmatrix} 0 & y \\ 0 & 0 \end{pmatrix} \right\}$$

are bi-ideals of S.

(2) Let

$$S = \left\{ \begin{pmatrix} a & b \\ c & d \end{pmatrix} | a, b, c, d \in \mathbb{N} \right\}.$$

Then, S is a semigroup under usual multiplication of matrices. The set

$$Q = \left\{ \begin{pmatrix} x & 0 \\ 0 & 0 \end{pmatrix} | x \in \mathbb{N} \right\}$$

is a quasi-ideal of S.

(3) Let $S = \{a, b, c, d, f\}$ be a semigroup with the following multiplication table.

$\cdot$	a	b	c	d	f
a	a	a	a	a	a
b	a	b	a	d	a
c	a	f	c	c	f
d	a	b	d	d	b
f	a	f	a	c	a

Quasi-ideals of S are

$$\{a\},\ \{a,\ b\},\ \{a,\ c\},\ \{a,\ d\},\ \{a,\ f\},\ \{a,\ b,\ c\}$$

$$\{a, c, d\},\ \{a,\ b,\ f\},\ \{a, c,\ f\},\ S.$$

Theorem 1.4.14. *Let $(S, \cdot)$ be a semigroup. If L is a left ideal and R is a right ideal of S, then the product $R \cdot L$ is a bi-ideal of S.*

Proof. Since $(R \cdot L) \cdot (R \cdot L) \subseteq R \cdot L$, the product $R \cdot L$ is a subsemigroup of S. On the other hand,

$$(R \cdot L) \cdot S \cdot (R \cdot L) \subseteq R \cdot S \cdot L \subseteq R \cdot L,$$

ie, the product $R \cdot L$ is a bi-ideal of S, as we stated. □

If S is a regular semigroup, then the converse of Theorem 1.4.14 also holds.

Theorem 1.4.15. *A subset A of a regular semigroup $(S, \cdot)$ is a bi-ideal of S if and only if S contains a left ideal L and a right ideal R such that $A = R \cdot L$.*

Proof. We prove that if S is a regular semigroup and A is a bi-ideal of S, then

$$A = (A \cup A \cdot S) \cdot (A \cup S \cdot A).$$

First, we see that

$$A \subseteq (A \cup A \cdot S) \cdot (A \cup S \cdot A) = A^2 \cup A \cdot S \cdot A,$$

because $a = a \cdot x \cdot a \in A \cdot S \cdot A$ for each $a \in A$. Conversely, $(A \cup A \cdot S) \cdot (A \cup S \cdot A) \subseteq A$, because A is a bi-ideal of S, ie, $A^2 \subseteq A$ and $A \cdot S \cdot A \subseteq A$. Thus, in view of Theorem 1.4.14, the theorem is proved. □

A more general result as that of Theorem 1.4.14 is contained in the following theorem.

Theorem 1.4.16. *The product of a bi-ideal and a non-empty subset of a semigroup S is also a bi-ideal of S.*

Proof. Let $(S, \cdot)$ be a semigroup, A be a non-empty subset, and B a bi-ideal of S. Then,

$$(A \cdot B) \cdot (A \cdot B) \subseteq A \cdot B,$$

ie, the product $A \cdot B$ is a subsemigroup of S. On the other hand,

$$(A \cdot B) \cdot S \cdot (A \cdot B) \subseteq A \cdot (B \cdot S \cdot B) \subseteq A \cdot B,$$

which shows that $A \cdot B$ is a bi-ideal of the semigroup S. □

Theorem 1.4.17. *For a semigroup $(S, \cdot)$ the following conditions are pairwise equivalent.*

(1) *S is a regular semigroup whose left ideals are two-sided.*

(2) *$B \cap L = B \cdot L$ for every bi-ideal B and every left ideal L of S.*

(3) *$L \cap Q = Q \cdot L$ for each left ideal L and each quasi-ideal Q of S.*

Proof. ($1 \Rightarrow 2$): Suppose that S is a regular semigroup whose left ideals are two-sided. Then, by Theorem 1.4.15, every bi-ideal B of S may be represented in the form $B = R \cdot I$, where R is a suitable right ideal and I is a suitable left ideal of S. Now, we obtain

$$B \cap L = R \cdot I \cap L = R \cdot I \cdot L = B \cdot L,$$

for every bi-ideal B and every left ideal L of S.

($2 \Rightarrow 3$): This is evident because every quasi-ideal of an arbitrary semigroup S is a bi-ideal of S.

($3 \Rightarrow 1$): Let S be a semigroup with property (3). Then, in the case $Q = R$, R is an arbitrary right ideal of S, and (3) implies that S is regular. Secondly, in the case $L = S$, $Q = L$, L is an arbitrary left ideal of S, and condition (3) implies

$$L = L \cap S = L \cdot S,$$

ie, any left ideal L is also a right ideal of S. The proof is completed. □

We state the left-right dual of Theorem 1.4.17.

Theorem 1.4.18. *For a semigroup $(S, \cdot)$ the following assertions are mutually equivalent.*

(1) *S is regular and each right ideal of S is two-sided.*

(2) *$B \cap R = R \cdot B$ for any bi-ideal B and for any right ideal R of S.*

(3) *$Q \cap R = R \cdot Q$ for every right ideal R and every quasi-ideal Q.*

1.5 HOMOMORPHISMS

In the construction of the theory of semigroups, it is natural to distinguish mapping of one semigroup into another under which the relations of the operation are preserved. A semigroup homomorphism is a function that preserves semigroup structure. In this section, we investigate homomorphism between semigroups.

Definition 1.5.1. If $(S, \cdot)$ and $(S', \star)$ are semigroups, then a map $f : S \to S'$ is a *homomorphism of semigroups* if $f(s_1 \cdot s_2) = f(s_1) \star f(s_2)$, for all $s_1, s_2 \in S$. If S has an identity, we require S' to have an identity and that $f(1) = 1$.

A homomorphism $f : S \to S'$ is

(1) a *monomorphism* if it is one-to-one function;

(2) an *epimorphism* if it is onto function;

(3) an *isomorphism* if it is both one-to-one and onto function.

When f is an isomorphism, we say S and S' are *isomorphic* and we write $S \cong S'$.

Example 15

(1) Consider the semigroups $(\mathbb{N}, +)$ and $(\mathbb{N}, \cdot)$, and define $f(n) = 2^n$, for all $n \in \mathbb{N}$. Then, $f(n+m) = 2^{n+m} = 2^n \cdot 2^m = f(n) \cdot f(m)$. Hence, f is a homomorphism.

(2) Consider the bicyclic semigroup defined in Example 1(10). The mapping $f : B \to \mathbb{Z}$ given by $f(a, b) = a - b$ is a homomorphism. Indeed, we have

$$\begin{aligned} f(a,b) + f(c,d) &= (a-b) + (c-d) \\ &= a - b + \max\{b,c\} - (d - c + \max\{b,c\}) \\ &= f(a - b + \max\{b,c\}, d - c + \max\{b,c\}) \\ &= f((a,b) \star (c,d)). \end{aligned}$$

Lemma 1.5.2. *Let $(S, \cdot)$, $(S', \star)$, and $(S'', *)$ be semigroups and $f : S \to S'$ and $g : S' \to S''$ be homomorphisms. Then, $g \circ f : S \to S''$ is a homomorphism.*

Proof. It is straightforward. □

A homomorphism of a semigroup into itself is called an *endomorphism*, while an isomorphism onto itself is called an *automorphism*.

Theorem 1.5.3. *The set of all semigroup endomorphisms of a semigroup is a semigroup under the operation of composition.*

Proof. From Lemma 1.5.2, it is clear. □

Theorem 1.5.3 can also be extended to automorphisms. The identity mapping is the identity with respect to composition of functions. Hence, the set of all the automorphisms of a semigroup is a monoid.

Proposition 1.5.4. *Let $(S, \cdot)$ be a semigroup. Then, there exists a homomorphism $f : S \to \mathcal{T}_S$, where $(\mathcal{T}_S, \circ)$ is the semigroup of functions from S to S under the operation of composition.*

Proof. We define $f : S \to \mathcal{T}_S$ by $f(a) = h_a$, for all $a \in S$, where $h_a \in \mathcal{T}_S$ and h_a is determined by $h_a(b) = a \cdot b$, for all $b \in S$. Now, we have $f(a \cdot b) = h_{a \cdot b}$, where

$$h_{a \cdot b}(c) = (a \cdot b) \cdot c = a \cdot (b \cdot c) = h_a(b \cdot c) = h_a(h_b(c)) = (h_a \circ h_b)(c).$$

Therefore, we obtain

$$f(a \cdot b) = h_{a \cdot b} = h_a \circ h_b = f(a) \circ f(b).$$

The last step shows that $f : S \to \mathcal{T}_S$ is a homomorphism from $(S, \cdot)$ into $(\mathcal{T}_S, \circ)$. □

Corollary 1.5.5. *Let $(S, \cdot)$ be a monoid. Then, there exists a subset T of $\mathcal{T}_S$ such that $(S, \cdot)$ is isomorphic to the monoid $(T, \circ)$.*

Proof. It is straightforward. □

Proposition 1.5.6. *Let $(S, \cdot)$ and $(S', \star)$ be semigroups. If $f : S \to S'$ is an isomorphism, then also the inverse map $f^{-1} : S' \to S$ is an isomorphism.*

Proof. Since f is one-to-one and onto, it follows that f^{-1} exists. Moreover, for all $x, y \in S'$, we have

$$f\left(f^{-1}(x) \cdot f^{-1}(y)\right) = f\left(f^{-1}(x)\right) \star f\left(f^{-1}(y)\right) = x \star y,$$

and so $f^{-1}(x \star y) = f^{-1}(x) \cdot f^{-1}(y)$, which proves f^{-1} is a homomorphism. □

We note some simple connections between homomorphisms, subsemigroups, and generating sets.

Theorem 1.5.7. *Let $(S, \cdot)$ and $(S', \star)$ be semigroups.*

(1) *Assume that $f : S \to S'$ is a homomorphism. Then, for any subsemigroup K of S, the image $f(K) = \{f(x) \mid x \in K\}$ of K under f is a subsemigroup of S'. Similarly, for any subsemigroup K' of S', the preimage $f^{-1}(K') = \{x \in S \mid f(x) \in K'\}$ of K' under f is a subsemigroup of S, if non-empty.*

(2) *For $f : S \to S'$ a homomorphism and $A \subseteq S$, f maps the subsemigroup of S generated by A onto the subsemigroup of S' generated by $f(A)$.*

(3) *Assume that A is a set of generators of S. Then, two homomorphisms f and g from S to S' coincide if and only if they coincide on A, ie, the equality $f(x) = g(x)$ holds for all $x \in A$.*

Proof. By straightforward calculations. In (3), consider the set $K = \{x \in S \mid f(x) = g(x)\}$. It is a subsemigroup of S and includes A. Hence, $S = \langle A \rangle \subseteq K$. □

1.6 CONGRUENCE RELATIONS AND ISOMORPHISM THEOREMS

The notion of congruence was first introduced by K. F. Gauss in the beginning of the 19th century. Congruence is a special type of equivalence relation which plays a vital role in the study of quotient structures of different algebraic structures. In this section, we study the quotient structure of semigroups by using the notion of congruence in semigroups.

Definition 1.6.1. Let ρ be a relation on a semigroup S.

(1) ρ is said to be *left compatible* if $(a, b) \in \rho$ and $x \in S$ imply $(x \cdot a, x \cdot b) \in \rho$;

(2) ρ is said to be *right compatible* if $(a, b) \in \rho$ and $x \in S$ imply $(a \cdot x, b \cdot x) \in \rho$;

(3) ρ is said to be *compatible* if it is both left and right compatible.

A right (left) compatible equivalence relation is called a *right (left) congruence*. A compatible equivalence relation on a semigroup S is called a *congruence*.

Lemma 1.6.2. *An equivalence relation ρ on a semigroup S is a congruence if and only if $(a, b) \in \rho$ and $(c, d) \in \rho$ imply $(a \cdot c, b \cdot d) \in \rho$.*

Proof. Suppose that ρ is a congruence. If $(a, b) \in \rho$ and $(c, d) \in \rho$, then by definition $(a \cdot c, a \cdot d) \in \rho$ and $(a \cdot d, b \cdot d) \in \rho$, and so $(a \cdot c, b \cdot d) \in \rho$ by transitivity. In the converse direction, the claim is trivial. □

We often write $x\rho y$ instead of $(x, y) \in \rho$.

Example 16. Consider the relation ρ on the bicyclic semigroup defined in Example 1(10), given by

$$(a, b)\rho(c, d) \Leftrightarrow a - b = c - d.$$

It is easy to verify that ρ is an equivalence relation. Further, suppose that $(a, b)\rho(c, d)$ and $(s, t)\rho(u, v)$. Then, $a - b = c - d$ and $s - t = u - v$. Thus,

$$a - b - (t - s) = c - d - (v - u).$$

We put $m = \max\{b, s\}$ and $n = \max\{d, u\}$. Then, we have

$$a - b + m - (t - s + m) = c - d + n - (v - u + n).$$

So, we obtain

$$(a - b + m, t - s + m)\rho(c - d + n, v - u + n).$$

This implies that

$$(a, b) \star (s, t)\rho(c, d) \star (u, v).$$

Therefore, ρ is a congruence on B.

Theorem 1.6.3. *Let $(S, \cdot)$ and $(S', \star)$ be semigroups and $f : S \to S'$ be a homomorphism. Corresponding to the homomorphism f, there exists a congruence ρ on S defined by*

$$x\rho y \Leftrightarrow f(x) = f(y), \text{ for } x, y \in S.$$

Proof. It is easy to see that ρ is an equivalence relation on S. Let $a, b, c, d \in S$ such that $a\rho b$ and $c\rho d$. Since

$$f(a \cdot c) = f(a) \star f(c) = f(b) \star f(d) = f(b \cdot d),$$

it follows that ρ is a congruence on S. □

The congruence relation defined in Theorem 1.6.3 is called the *kernel* of f, ie,

$$kerf = \{(x, y) \in S \times S | f(x) = f(y)\}.$$

Corollary 1.6.4. *For each homomorphism $f : S \to S'$, $kerf$ is a congruence of S.*

We prove now that the congruences of a semigroup $(S, \cdot)$ are closed under intersections.

Proposition 1.6.5. *Let $(S, \cdot)$ be a semigroup and $\{\rho_i\}_{i\in\Lambda}$ be a family of congruences of S. Then, $\bigcap_{i\in\Lambda} \rho_i$ is a congruence of S too.*

Proof. Denote $\rho = \bigcap_{i\in\Lambda} \rho_i$. Suppose that $(a, b) \in \rho$ and $c \in S$. Then, for all $i \in \Lambda$, $(a, b) \in \rho_i$, and so $(a \cdot c, b \cdot c) \in \rho_i$ and $(c \cdot a, c \cdot b) \in \rho_i$. This implies that $(a \cdot c, b \cdot c) \in \rho$ and $(c \cdot a, c \cdot b) \in \rho$. Therefore, ρ is a congruence of S. □

Let $(S, \cdot)$ be a semigroup and ρ be a congruence relation on S. We denote $\rho(a)$ the equivalence class of a. We set

$$S/\rho = \{\rho(a) \,| a \in S\}.$$

Theorem 1.6.6. *Let $(S, \cdot)$ be a semigroup and ρ be a congruence relation on S. The quotient set S/ρ together with the following binary operation*

$$\rho(a) \odot \rho(b) = \rho(a \cdot b)$$

is a semigroup. If $(S, \cdot)$ is a monoid, then so is $(S/\rho, \odot)$.

Proof. We need to make sure that this operation is well-defined. If $\rho(a) = \rho(a')$ and $\rho(b) = \rho(b')$, then $a\rho a'$ and $b\rho b'$. Since ρ is a congruence relation, it follows that $a \cdot b\rho a' \cdot b'$ and so $\rho(a \cdot b) = \rho(a' \cdot b')$. Thus, the binary operation is well-defined. The associativity of the operation $\cdot$ guarantees the associativity of the operation $\odot$ on S/ρ. Indeed, for $\rho(a), \rho(b), \rho(c) \in S/\rho$, we have

$$\rho(a) \odot (\rho(b) \odot \rho(c)) = \rho(a) \odot \rho(b \cdot c) = \rho(a \cdot (b \cdot c)) = \rho((a \cdot b) \cdot c)$$

$$= \rho(a \cdot b) \odot \rho(c) = (\rho(a) \odot \rho(b)) \odot \rho(c).$$

Now, suppose that S is a monoid. Then, we have

$$\rho(1) \odot \rho(a) = \rho(1 \cdot a) = \rho(a) = \rho(a \cdot 1) = \rho(a) \odot \rho(1)$$

for all $a \in S$. Hence, we conclude that $(S/\rho, \odot)$ is a semigroup and if $(S, \cdot)$ is a monoid, then so is $(S/\rho, \odot)$. □

We call S/ρ the *quotient semigroup* (or *monoid*) of S by ρ.

Corollary 1.6.7. *There exists a homomorphism from $(S, \cdot)$ to $(S/\rho, \odot)$ called natural homomorphism.*

Proof. We define a mapping $\varphi : S \to S/\rho$ by $\varphi(a) = \rho(a)$, for all $a \in S$. Then, for $a, b \in S$,

$$\varphi(a \cdot b) = \rho(a \cdot b) = \rho(a) \odot \rho(b) = \varphi(a) \odot \varphi(b).$$

So, φ is a homomorphism. □

Now, we want to examine the kernel of φ. Indeed, we have

$$\begin{aligned} ker\varphi &= \{(x, y) \in S \times S | \varphi(x) = \varphi(y)\} \\ &= \{(x, y) \in S \times S | \rho(x) = \rho(y)\} \\ &= \{(x, y) \in S \times S | (x, y) \in \rho\} \\ &= \rho. \end{aligned}$$

Therefore, we have the following corollary.

Corollary 1.6.8. *Every congruence is a kernel of some homomorphism.*

Theorem 1.6.9. *Let $(S, \cdot)$ and $(S', \star)$ be semigroups and $f : S \to S'$ be a homomorphism. Then, there exists a unique monomorphism $g : S/kerf \to S'$ such that the following diagram commutes, ie, $f = g \circ \varphi$.*

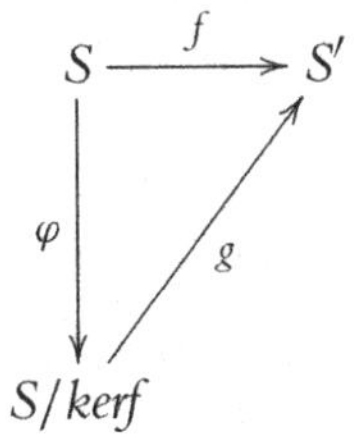

Proof. Suppose that $\rho = kerf$ and $\varphi : S \to S/\rho$ the corresponding natural homomorphism. Define $g : S/\rho \to S'$ by $g(\rho(x)) = f(x)$, for all $x \in S$. First, we show that g is well-defined. Suppose that $\rho(x) = \rho(y)$. Then, $(x, y) \in kerf$ and so $f(x) = f(y)$. This implies that $g(\rho(x)) = g(\rho(y))$. Thus, g is well-defined. Now, for $\rho(x), \rho(y) \in S/\rho$, we have

$$\begin{aligned} g(\rho(x) \odot \rho(y)) &= g(\rho(x \cdot y)) = f(x \cdot y) \\ &= f(x) \star f(y) = g(\rho(x)) \star g(\rho(y)). \end{aligned}$$

Thus, g is a homomorphism. It is straightforward to see that g is one-to-one. Finally, g is unique. If $h : S/\rho \rightarrow S'$ is another such one-to-one homomorphism, then $f = h \circ \varphi$. Hence, $f(x) = h(\rho(x))$, for all $x \in S$. But this means that $g = h$. □

Theorem 1.6.10 (First Isomorphism Theorem). *Let $(S, \cdot)$ and $(S', \star)$ be semigroups and $f : S \rightarrow S'$ be a homomorphism. Then,*

$$S/kerf \cong f(S).$$

Proof. The homomorphism f is onto as a homomorphism $S \rightarrow f(S)$. When Theorem 1.6.9 is applied to $f : S \rightarrow f(S)$ we obtain a unique monomorphism $g : S/kerf \rightarrow f(S)$. Moreover, g is onto. Consequently, g is an isomorphism. □

Remark 3. Note that the analogous result holds for monoids.

Suppose that we have two congruences σ and ρ on a semigroup $(S, \cdot)$ such that σ is a subset of ρ. Then, there exists a unique homomorphism $\psi : S/\sigma \rightarrow S/\rho$ making the following diagram commute:

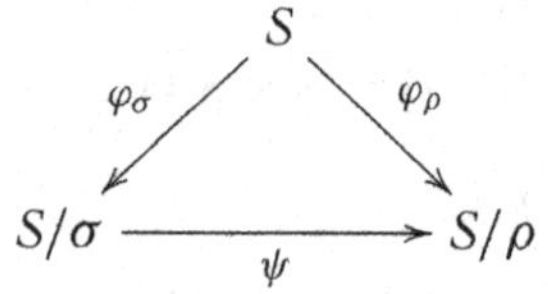

where we denote $ker\psi$ by ρ/σ and $(\sigma(a), \sigma(b)) \in \rho/\sigma$ if and only if $a\rho b$.

Theorem 1.6.11 (Second Isomorphism Theorem). *Let σ and ρ be two congruences on a semigroup $(S, \cdot)$ such that σ is a subset of ρ. Then,*

$$\rho/\sigma = \{(\sigma(a), \sigma(b)) \in S/\sigma \times S/\sigma \mid (a, b) \in \rho\}$$

is a congruence on S/σ and

$$(S/\sigma)/(\rho/\sigma) \cong S/\rho.$$

Proof. Define $f : S/\sigma \rightarrow S/\rho$ by $f(\sigma(x)) = \rho(x)$, for $x \in S$. It is easy to see that f is an onto homomorphism and $kerf = \rho/\sigma$. □

Let $(S, \cdot)$ be a semigroup and I be an ideal of S. We define the relation ρ_I as follows:

$$x\rho_I y \Leftrightarrow \text{ either } x = y \text{ or both } x \text{ and } y \text{ are in } I.$$

This means that ρ_I contacts the ideal I and leaves the rest of S as it was.

Proposition 1.6.12. *The relation ρ_I is a congruence relation on S.*

Proof. It is straightforward. □

The relation ρ_I is called *Rees congruence*. The quotient semigroup S/ρ_I is denoted simply S/I and it is called the *Rees quotient* (*of the ideal I*).

1.7 GREEN'S RELATIONS

Green's relations are five equivalences ($\mathcal{L}, \mathcal{R}, \mathcal{J}, \mathcal{D}$, and $\mathcal{H}$) on a semigroup based on the idea of mutual divisibility of elements. They are important tools in the description and decomposition of semigroups. These relations were introduced explicitly in a paper by Green in 1951 [1].

Definition 1.7.1. Let $(S, \cdot)$ be a semigroup and define the following relations on S:

$$x\mathcal{L}y \Leftrightarrow S^1 \cdot x = S^1 \cdot y,$$
$$x\mathcal{R}y \Leftrightarrow x \cdot S^1 = y \cdot S^1,$$
$$x\mathcal{J}y \Leftrightarrow S^1 \cdot x \cdot S^1 = S^1 \cdot y \cdot S^1.$$

Here $S^1 \cdot x$, $x \cdot S^1$, and $S^1 \cdot x \cdot S^1$ are the principal left ideal, the principal right ideal, and the principal ideal generated by $x \in S$, respectively. By the definitions, we obtain

$$x\mathcal{L}y \Leftrightarrow \text{ there exist } s, s' \in S^1 \text{ such that } x = s \cdot y \text{ and } y = s' \cdot x,$$
$$x\mathcal{R}y \Leftrightarrow \text{ there exist } r, r' \in S^1 \text{ such that } x = y \cdot r \text{ and } y = x \cdot r'.$$

Lemma 1.7.2

(1) *$\mathcal{L}$ is a right congruence.*

(2) *$\mathcal{R}$ is a left congruence.*

Proof. It is straightforward. □

Remark 4. If $a\mathcal{L}b$, then $S^1 \cdot a = S^1 \cdot b$, which implies that $S^1 \cdot a \cdot S^1 = S^1 \cdot b \cdot S^1$ and so $a\mathcal{J}b$. Thus, $\mathcal{L} \subseteq \mathcal{J}$. Similarly, we have $\mathcal{R} \subseteq \mathcal{J}$.

If ρ and σ are relations on S, we define

$$\rho \circ \sigma = \{(x, y) \in S \times S | \exists z \in S \text{ with } (x, z) \in \rho \text{ and } (z, y) \in \sigma\}.$$

Lemma 1.7.3. *If ρ and σ are equivalence relations on S and if $\rho\circ\sigma = \sigma\circ\rho$, then $\rho \circ \sigma$ is an equivalence relation. Moreover, it is the smallest equivalence relation containing $\rho \cup \sigma$.*

Proof. Suppose that $\delta = \rho \circ \sigma = \sigma \circ \rho$.

For any $a \in S$, we have $a\rho a$ and $a\sigma a$. Hence $a\delta a$, which implies that δ is reflexive.

If $a\delta b$, then there exists $c \in S$ such that $a\rho c$ and $c\sigma b$. Since ρ and σ are symmetric, it follows that $b\sigma c$ and $c\rho a$. This implies that $b\delta a$, ie, δ is symmetric.

Now, assume that $a\delta b$ and $b\delta c$. Then, there exist $x, y \in S$ such that

$$a\rho x, x\sigma b \quad \text{and} \quad b\sigma y, y\rho c.$$

Since $x\sigma b$ and $b\sigma y$, it follows that $x\sigma y$. Thus, from $x\sigma y$ and $y\rho c$ we obtain $x\delta c$. Hence, there exists $z \in S$ such that $x\rho z$ and $z\sigma c$. Since $a\rho x$ and $x\rho z$, it follows that $a\rho z$. Therefore, we have $a\rho z$ and $z\sigma c$. This implies that $a\delta c$. Thus, δ is transitive. We have shown that δ is an equivalence relation.

Let $(a, b) \in \rho$. Since $a\rho b$ and $b\sigma b$, it follows that $(a, b) \in \delta$. Similarly, let $(a, b) \in \sigma$. Since $a\rho a$ and $a\sigma b$, it follows that $(a, b) \in \delta$. Thus, we conclude that $\rho \cup \sigma \subseteq \delta$.

Now, suppose that δ' is another equivalence relation such that $\rho \cup \sigma \subseteq \delta'$. Let $(a, b) \in \delta$. Then, there exists $c \in S$ such that $a\rho c$ and $c\sigma b$. Hence, $a\delta' c$ and $c\delta' b$. This implies that $a\delta' b$, since δ' is transitive. This means that $(a, b) \in \delta'$. Therefore, $\delta \subseteq \delta'$. □

Theorem 1.7.4. *The relations $\mathcal{L}$ and $\mathcal{R}$ commute, ie, $\mathcal{L} \circ \mathcal{R} = \mathcal{R} \circ \mathcal{L}$.*

Proof. Suppose that $(x, y) \in \mathcal{L} \circ \mathcal{R}$. This means that there exists an element $z \in S$ such that $x\mathcal{L}z$ and $z\mathcal{R}y$. Therefore, there are elements $s, s', r, r' \in S^1$ such that

$$x = s \cdot z, \quad z = s' \cdot x, \quad z = y \cdot r, \quad y = z \cdot r'.$$

Denote $t = s \cdot z \cdot r'$. Now, we have

$$t = s \cdot z \cdot r' = x \cdot r',$$

$$x = s \cdot z = s \cdot y \cdot r = s \cdot z \cdot r' \cdot r = t \cdot r,$$

which means that $x\mathcal{R}t$. On the other hand, we have

$$t = s \cdot z \cdot r' = s \cdot y,$$

$$y = z \cdot r' = s' \cdot x \cdot r' = s' \cdot s \cdot z \cdot r' = s' \cdot t,$$

which implies that $y\mathcal{L}t$. Since $x\mathcal{R}t$ and $t\mathcal{L}y$, it follows that $(x, y) \in \mathcal{R} \circ \mathcal{L}$. Thus, we have shown that $\mathcal{L} \circ \mathcal{R} \subseteq \mathcal{R} \circ \mathcal{L}$. The inclusion in the other direction is proved in the same way. □

Definition 1.7.5. We define

$$\mathcal{D} = \mathcal{L} \circ \mathcal{R} \quad \text{and} \quad \mathcal{H} = \mathcal{L} \cap \mathcal{R}.$$

Lemma 1.7.6. *$\mathcal{D}$ is the smallest equivalence relation that contains both $\mathcal{L}$ and $\mathcal{R}$.*

Proof. By Theorem 1.7.4 and Lemma 1.7.3. □

Lemma 1.7.7. *$\mathcal{H}$ is an equivalence relation, and it is the largest equivalence relation of S that is contained in both $\mathcal{L}$ and $\mathcal{R}$.*

Proof. It is straightforward. □

Now, by definition

$$\mathcal{H} = \mathcal{L} \cap \mathcal{R} \subseteq \mathcal{L} \subseteq \mathcal{D},$$

$$\mathcal{H} = \mathcal{L} \cap \mathcal{R} \subseteq \mathcal{R} \subseteq \mathcal{D}.$$

As $\mathcal{J}$ is an equivalence relation and $\mathcal{L} \cup \mathcal{R} \subseteq \mathcal{J}$, we must have

$$\mathcal{D} \subseteq \mathcal{J}.$$

The inclusion diagram of Green's relations is given in the following figure.

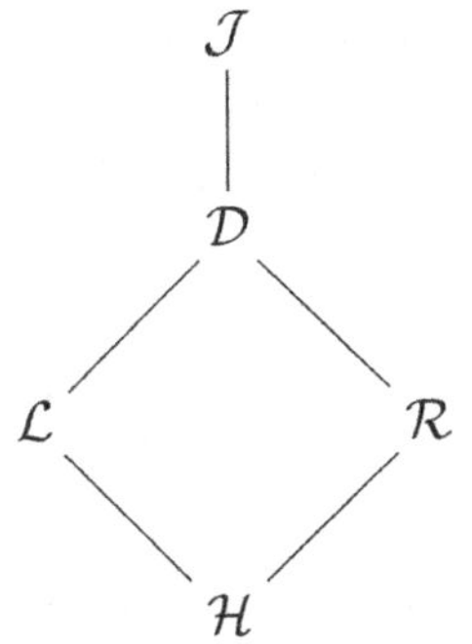

The corresponding equivalence classes containing $a \in S$ are denoted by $\mathcal{L}_a$, $\mathcal{R}_a$, $\mathcal{J}_a$, $\mathcal{D}_a$, and $\mathcal{H}_a$, respectively.

Remark 5. Clearly, for all $a \in S$, $\mathcal{H}_a = \mathcal{L}_a \cap \mathcal{R}_a$.

Proposition 1.7.8. *Let $(S, \cdot)$ be a semigroup. Then,*

$$x\mathcal{D}y \Leftrightarrow \mathcal{L}_x \cap \mathcal{R}_y \neq \emptyset \Leftrightarrow \mathcal{L}_y \cap \mathcal{R}_x \neq \emptyset.$$

Moreover,

$$\mathcal{D}_x = \bigcup_{y \in \mathcal{D}_x} \mathcal{L}_y = \bigcup_{y \in \mathcal{D}_x} \mathcal{R}_y.$$

Proof. By the definition of $\mathcal{D}$, $x\mathcal{D}y \Leftrightarrow \exists z \in S$ such that $x\mathcal{L}z$ and $z\mathcal{R}y \Leftrightarrow \exists s \in S$ such that $x\mathcal{R}s$ and $s\mathcal{L}y$. The first claim follows from this. The second claim is trivial, since $\mathcal{L} \subseteq \mathcal{D}$ and $\mathcal{R} \subseteq \mathcal{D}$. □

1.8 FREE SEMIGROUPS

In this section, we define a very special example of semigroups: the free ones.

Let A be a set of symbols, called an *alphabet*. Its elements are *letters*. Any finite sequence of letters is a *word* over A.

We denote by A^+ the set of all ordered sequences of finite length greater than 0, ie, a typical element of A^+ has the form

$$w = (a_1, \ldots, a_n),$$

where $n \in \mathbb{N}$ and $a_1, \ldots, a_n$ are any elements of A. We usually omit the brackets and the periods in the notation of w, thus abbreviating

$$w = a_1, \ldots, a_n.$$

Two words $w = a_1, \ldots, a_n$ and $v = b_1, \ldots, b_m$ from A^+ coincide if their lengths m and n are the same and $a_i = b_i$ holds for all $1 \leq i \leq n$. The set A^+ is a semigroup, the *free semigroup* over A, when the product is defined as the *catenation* of words, ie, the product of the words $w_1 = a_1, \ldots, a_n$ and $w_2 = b_1, \ldots, b_m$ $(a_i, b_i \in A)$ is the word $w_1 \star w_2 = a_1, \ldots, a_n b_1, \ldots, b_m$. When we join the empty word 1 (which has no letters) into A^+, we have the *free monoid* A^*, $A^* = A^+ \cup \{1\}$. Clearly, $1 \star w = w = w \star 1$, for all words $w \in A$.

Theorem 1.8.1. *Let $(A^+, \star)$ be the free semigroup over A, $(S, \cdot)$ be an arbitrary semigroup, and $f : A \to S$ be an arbitrary map. Then, there is a unique homomorphism $\bar{f} : A^+ \to S$ extending f.*

Proof. Suppose that $(S, \cdot)$ is a semigroup. Every word $w = a_1, \ldots, a_n$ over A is the concatenation of its one-letter subwords $a_1, \ldots, a_n$. It follows that for a homomorphism $g : A^+ \to S$ extending f, we must have $g(w) = f(a_1) \cdot \cdots \cdot f(a_n)$, which proves the uniqueness statement.

On the other hand, the map $\bar{f} : A^+ \to S$ given by $\bar{f}(w) = f(a_1) \cdot \cdots \cdot f(a_n)$ is a homomorphism from A^+ into S which extend f. □

Remark 6. Theorem 1.8.1 is the reason for calling the semigroup A^+ free over A. The elements of A^+ do not satisfy any equations, except those that hold in all semigroups.

Example 17. The free semigroup over a one-letter alphabet $A = \{a\}$ is isomorphic to $(\mathbb{N}^*, +)$, the additive semigroup of positive natural numbers. The words over A have the form $w = a^n$, where $n \in \mathbb{N}^*$ and a^n is the sequence of length n with all entries equal to a. The mapping $f : \mathbb{N}^* \to \{a\}^+$ defined by $n \mapsto a^n$, gives us an isomorphism from $(\mathbb{N}^*, +)$ onto $(\{a\}^+, \star)$.

Corollary 1.8.2. *Every semigroup is the homomorphic image of a free one.*

Proof. Let $(S, \cdot)$ be a semigroup. We consider an alphabet A of size at least $|S|$ and a map f from A onto S. The homomorphic extension $\overline{f} : A^+ \to S$ of f, given by Theorem 1.8.1, maps A^+ onto S. □

Corollary 1.8.3. *If A^+ and B^+ are free semigroups such that $|A| = |B|$, then $A^+ \cong B^+$.*

Proof. It is straightforward. □

Corollary 1.8.4. *Every free semigroup is cancellative.*

Proof. It is straightforward. □

1.9 APPROXIMATIONS IN A SEMIGROUP

Let $(S, \cdot)$ be a semigroup. We consider the relation α and its transitive closure α^*. The relation α is the smallest equivalence relation on S so that S/α^* is a commutative semigroup. This relation is studied by Freni [19]. Based on the relation α, we define a neighborhood system for each element of S, and we present a general framework of the study of approximations in semigroups. The connections between semigroups and operators are examined. The main reference for this section is [20, 21].

Definition 1.9.1. If $(S, \cdot)$ is a semigroup, then we set: $\alpha_1 = \{(x, x) \mid x \in S\}$ and, for every integer $n > 1$, α_n is the relation defined as follows:

$$x\alpha_n y \Leftrightarrow \exists(z_1, z_2, \ldots, z_n) \in S^n, \exists\sigma \in \mathbb{S}_n : x = \prod_{i=1}^{n} z_i, y = \prod_{i=1}^{n} z_{\sigma(i)},$$

where $\mathbb{S}_n$ is the symmetric group of order n.

Obviously, for every $n \geq 1$, the relations α_n are symmetric, and the relation $\alpha = \bigcup_{n\geq 1} \alpha_n$ is reflexive and symmetric. Let α^* be the *transitive closure* of α.

In [19], it is proved that α^* is the smallest congruence of S such that the quotient S/α^* is a commutative semigroup. We shall explain more about this relation in future.

Let η^* be a congruence on S such that S/η^* is a commutative semigroup. Clearly, $\alpha^* \subseteq \eta^*$. We put $\alpha^*/\eta^* = \{(\eta^*(x), \eta^*(y)) \in S/\eta^* \times S/\eta^* \mid (x, y) \in \alpha^*\}$, then α^*/η^* is a congruence on S/η^*.4

Let $(S, \cdot)$ be a semigroup. For the relation α on S and a positive integer k, we now define a notion of binary relation α^k called the *k-step-relation* of α as follows:

(1) $\alpha^1 = \alpha$;

(2) $\alpha^k = \{(x, y) \in S \times S|$ there exist $y_1, y_2, \ldots, y_i \in S,\ 1 \leq i \leq k-1$, such that $x\alpha y_1, y_1\alpha y_2, \ldots, y_i\alpha y\} \cup \alpha^1$, $k \geq 2$.

It is easy to see that

$$\alpha^{k+1} = \alpha^k \cup \{(x, y) \in S \times S| \text{ there exist } y_1, \ldots, y_k \in S, \text{ such that } x\alpha y_1,\ y_1\alpha y_2, \ldots, y_k\alpha y\}.$$

Obviously, $\alpha^k \subseteq \alpha^{k+1}$, and there exists $n \in \mathbb{N}$ such that $\alpha^k = \alpha^n$ for all $k \geq n$. (In fact $\alpha^n = \alpha^*$ is nothing else but the transitive closure of α). Of course α^* is transitive. The relation α^k can be conveniently expressed as a mapping from S to $\wp(S)$, $N_k(x) = \{y \in S\ |x\alpha^k y\}$ by collecting all α^k-related elements for each element $x \in S$. The set $N_k(x)$ may be viewed as a α^k*-neighborhood* of x defined by the binary relation α^k.

Based on the relation α^k on S, we can obtain a *neighborhood system* for each element x: $\{N_k(x)|k \geq 1\}$. This neighborhood system is monotonically increasing with respect to k. We can also observe that

$$N_k(x) = \{y \in S| \exists y_1, y_2, \ldots, y_i \in S \text{ such that } x\alpha y_1,\ y_1\alpha y_2, \ldots, y_i\alpha y,\quad 1 \leq i \leq k-1, \text{ or } x\alpha y\}.$$

Theorem 1.9.2. *Let $(S, \cdot)$ be a semigroup. For each $a, b \in S$ and natural numbers k, l we have*

$$N_k(a) \cdot N_l(b) \subseteq N_{k+l-1}(a \cdot b).$$

Proof. Suppose that $x \in N_k(a) \cdot N_l(b)$. Then, there exist $a' \in N_k(a)$ and $b' \in N_l(b)$ such that $x = a' \cdot b'$. Since $a' \in N_k(a)$, it follows that $a'\alpha^k a$ and so there exist $\{x_1, \ldots, x_{k+1}\} \subseteq S$ with $x_1 = a'$, $x_{k+1} = a$ and permutations σ_i such that

$$x_i = \prod_{j=1}^{n_i} u_{ij}, \quad x_{i+1} = \prod_{j=1}^{n_i} u_{i\sigma_i(j)},$$

$i = 1, \ldots, k$; and since $b' \in N_l(b)$, it follows that $b'\alpha^l b$ and so there exist $\{y_1, \ldots, y_{l+1}\} \subseteq S$ with $y_1 = b'$, $y_{l+1} = b$ and permutations τ_j such that

$$y_j = \prod_{r=1}^{m_j} v_{rj}, \quad y_{j+1} = \prod_{r=1}^{m_j} v_{\tau_j(r)j},$$

$j = 1, \ldots, l$. Therefore, we obtain

$$x_i \cdot y_i = \prod_{j=1}^{n_i} u_{ij} \prod_{r=1}^{m_i} v_{ri}, \quad x_{i+1} \cdot y_i = \prod_{j=1}^{n_i} u_{i\sigma_i(j)} \prod_{r=1}^{m_i} v_{ri}$$

and

$$x_{k+1} \cdot y_j = \prod_{j=1}^{n_k} u_{k\sigma_k(j)} \prod_{r=1}^{m_j} v_{rj}, \quad x_{k+1} \cdot y_{j+1} = \prod_{j=1}^{n_k} u_{k\sigma_k(j)} \prod_{r=1}^{m_j} v_{\tau_j(r)j}.$$

If we pick up elements $z_1, \ldots, z_{k+l}$ such that

$$z_i = x_i \cdot y_i, \quad i = 1, \ldots, k,$$
$$z_{k+j} = x_{k+1} \cdot y_{j+1}, \quad j = 1, \ldots, l,$$

then, $z_m \ \alpha^{k+l-1} \ z_{m+1}$, $m = 1, \ldots, k+l-1$. So, we have $x = a' \cdot b' = x_1 \cdot y_1 \ \alpha^{k+l-1} \ x_{k+1} \cdot y_{l+1} = a \cdot b$. Therefore, $x \in N_{k+l-1}(a \cdot b)$. □

For a neighborhood operator N_k on S, we can extend N_k from $\mathcal{P}^*(S)$ to $\mathcal{P}^*(S)$ by: $N_k(X) = \bigcup_{x \in X} N_k(x)$ for all $X \subseteq S$. So we can directly deduce that:

Proposition 1.9.3

(1) $A \subseteq B \Rightarrow N_k(A) \subseteq N_k(B)$;

(2) *For all* $k, l \geq 1$, *we have* $N_l(N_k(x)) \subseteq N_{l+k}(x)$.

Definition 1.9.4. Let $(S, \cdot)$ be a semigroup and A be a non-empty subset of S. We define the *lower and upper approximations* of A with respect to α^* as follows:

$$\underline{\alpha^*(A)} = \{x \in S | \alpha^*(x) \subseteq A\} \quad \text{and} \quad \overline{\alpha^*(A)} = \{x \in S | \alpha^*(x) \cap A \neq \emptyset\}.$$

Similarly, we can define the lower and upper approximations of A with respect to η^*. In this case, we have

$$\underline{\eta^*(A)} \subseteq \underline{\alpha^*(A)} \subseteq A \subseteq \overline{\alpha^*(A)} \subseteq \overline{\eta^*(A)}.$$

In [22], Kuroki gave some properties of the lower and upper approximations with respect to the congruence relations; see also [23]. Since α^* is a congruence relation, all the results in [22] are true for the relation α^*.

Definition 1.9.5. For the relation α, by substituting equivalence class $\alpha^*(x)$ with α^k-neighborhood $N_k(x)$ in the previous definition, we can define a pair of lower and upper *approximation operators* with respect to N_k as follows:

$$\underline{apr}_k(A) = \{x \in S | N_k(x) \subseteq A\} \quad \text{and} \quad \overline{apr}_k(A) = \{x \in S | N_k(x) \cap A \neq \emptyset\}.$$

The set $\underline{apr}_k(A)$ consists of those elements whose α^k-neighborhoods are contained in A, and $\overline{apr}_k(A)$ consists of those elements whose α^k-neighborhoods have a non-empty intersection with A.

Proposition 1.9.6. *If A is a non-empty subset of a semigroup $(S, \cdot)$, then we have*

(1) $\underline{apr}_{k+1}(A) \subseteq \underline{apr}_k(A)$;
(2) $\overline{apr}_k(A) \subseteq \overline{apr}_{k+1}(A)$.

Therefore,

Corollary 1.9.7. *We have*

$$\bigcup\{x|x \in \underline{\alpha^*(A)}\} = \bigcap_k \underline{apr}_k(A) \text{ and } \bigcup\{x|x \in \overline{\alpha^*(A)}\} = \bigcup_k \overline{apr}_k(A).$$

Proposition 1.9.8. *If A and B are non-empty subsets of a semigroup $(S, \cdot)$, then the pair of approximation operators satisfies the following properties:*

(1) $\underline{apr}_k(A) \subseteq A \subseteq \overline{apr}_k(A)$;
(2) $\underline{apr}_k(A) = (\overline{apr}_k(A^c))^c$;
(3) $\overline{apr}_k(A) = (\underline{apr}_k(A^c))^c$;
(4) $\underline{apr}_k(A \cap B) = \underline{apr}_k(A) \cap \underline{apr}_k(B)$;
(5) $\overline{apr}_k(A \cup B) = \overline{apr}_k(A) \cup \overline{apr}_k(B)$;
(6) $\underline{apr}_k(A \cup B) \supseteq \underline{apr}_k(A) \cup \underline{apr}_k(B)$;
(7) $\overline{apr}_k(A \cap B) \subseteq \overline{apr}_k(A) \cap \overline{apr}_k(B)$;
(8) $A \subseteq B \Longrightarrow \underline{apr}_k(A) \subseteq \underline{apr}_k(B)$;
(9) $A \subseteq B \Longrightarrow \overline{apr}_k(A) \subseteq \overline{apr}_k(B)$.

Proof. It is straightforward. □

Theorem 1.9.9. *Let A be a non-empty subset of a semigroup $(S, \cdot)$. For all $k \geq l \geq 1$, we have*

(1) $A \subseteq \underline{apr}_l(\overline{apr}_k(A))$;
(2) $\overline{apr}_l(\underline{apr}_k(A)) \subseteq A$.

Proof. (1) Suppose that $a \in A$. If $N_l(a) = \emptyset$. Then it is clear that $N_l(a) \subseteq \overline{apr}_k(A)$, which implies that $a \in \underline{apr}_l(\overline{apr}_k(A))$, and so $A \subseteq \underline{apr}_l(\overline{apr}_k(A))$. If $N_l(a) \neq \emptyset$, then for each $b \in N_l(a)$, we have $a \in N_l(b)$. Hence $N_l(b) \cap A \neq \emptyset$. Now, we have $b \in \overline{apr}_l(A)$, and by Proposition 1.9.6, we obtain $b \in \overline{apr}_k(A)$. Therefore, $N_l(a) \subseteq \overline{apr}_k(A)$, which implies that $a \in \underline{apr}_l(\overline{apr}_k(A))$, and so $A \subseteq \underline{apr}_l(\overline{apr}_k(A))$.

(2) Suppose that $a \in \overline{apr}_l(\underline{apr}_k(A))$. Then, we have $N_l(a) \cap \underline{apr}_k(A) \neq \emptyset$, and so there exists $b \in N_l(a) \cap \underline{apr}_k(A)$. Therefore, $a \in N_l(b)$ and $N_k(b) \subseteq A$. Hence, $a \in N_l(b) \subseteq N_k(b) \subseteq A$, and so we conclude that $\overline{apr}_l(\underline{apr}_k(A)) \subseteq A$. □

Theorem 1.9.10. *For all $k, l \geq 1$ and $A \subseteq S$, we have*

(1) $\underline{apr}_{l+k}(A) \subseteq \underline{apr}_l(\underline{apr}_k(A))$;
(2) $\overline{apr}_{l+k}(A) \supseteq \overline{apr}_l(\overline{apr}_k(A))$.

Proof. (1) Suppose that $a \in \underline{apr}_{l+k}(A)$. Then, $N_{l+k}(a) \subseteq A$. Using Proposition 1.9.3, we have $N_k(N_l(a)) \subseteq N_{k+l}(a) \subseteq A$, which implies that $N_l(a) \subseteq \underline{apr}_k(A)$. Therefore, $a \in \underline{apr}_l(\underline{apr}_k(A))$.

(2) Suppose that $a \in \overline{apr}_l(\overline{apr}_k(A))$. Then $N_l(a) \cap \overline{apr}_k(A) \neq \emptyset$, and so there exists $b \in N_l(a) \cap \overline{apr}_k(A)$. Since $b \in \overline{apr}_k(A)$, it follows that $N_k(b) \cap A \neq \emptyset$. Now, we have

$$\emptyset \neq N_k(b) \cap A \subseteq N_k(N_l(a)) \cap A \subseteq N_{l+k}(a) \cap A,$$

and so $N_{l+k}(a) \cap A \neq \emptyset$, which implies that $a \in \overline{apr}_{l+k}(A)$. □

Theorem 1.9.11. *If A, B are non-empty subsets of a semigroup $(S, \cdot)$, then*

$$\overline{apr}_k(A) \cdot \overline{apr}_l(B) \subseteq \overline{apr}_{k+l-1}(A \cdot B).$$

Proof. Suppose that z be any element of $\overline{apr}_k(A) \cdot \overline{apr}_l(B)$. Then, there exist $x \in \overline{apr}_k(A)$ and $y \in \overline{apr}_l(B)$ such that $z = x \cdot y$. Since $x \in \overline{apr}_k(A)$ and $y \in \overline{apr}_l(B)$, it follows that there exist $a, b \in S$ such that $a \in N_k(x) \cap A$ and $b \in N_l(y) \cap B$. So $a \in N_k(x)$ and $b \in N_l(y)$. By Theorem 1.9.2, we have $N_k(x) \cdot N_l(y) \subseteq N_{k+l-1}(z)$. Since $a \cdot b \in A \cdot B$, we obtain $a \cdot b \in N_{k+l-1}(z) \cap A \cdot B$, and so $z \in \overline{apr}_{k+l-1}(A \cdot B)$. This completes the proof. □

Corollary 1.9.12. *Let $(S, \cdot)$ be a semigroup and A be a closed subset of S, ie, $A \cdot A \subseteq A$, then $\overline{apr}_1(A)$ is also a closed subset of S.*

Theorem 1.9.13. *Let $\alpha^n = \alpha^*$ for some $n \in \mathbb{N}$. If A is a subsemigroup of S, then $\overline{apr}_n(A)$ is also a subsemigroup of S.*

Proof. Suppose that $x, y \in \overline{apr}_n(A)$. Then $N_n(x) \cap A \neq \emptyset$ and $N_n(y) \cap A \neq \emptyset$. Hence, there exist $a \in N_n(x) \cap A$ and $b \in N_n(y) \cap A$. Since $N_n = N_k$ for all $k \geq n$, by Theorem 1.9.2, we obtain

$$a \cdot b \in N_n(x) \cdot N_n(y) \subseteq N_{2n-1}(x \cdot y) = N_n(x \cdot y).$$

On the other hand, since A is a subsemigroup, it follows that $a \cdot b \in A$. Thus $N_n(x \cdot y) \cap A \neq \emptyset$ which implies that $x \cdot y \in \overline{apr}_n(A)$. Therefore, for every $x \in \overline{apr}_n(A)$, we obtain $x \cdot \overline{apr}_n(A) \subseteq \overline{apr}_n(A)$. □

1.10 ORDERED SEMIGROUPS

Ordered semigroups have been studied by several authors, for example, Alimov [24], Clifford [25], and Kehayopulu et al. [26–32]. Saito [33] studied ordered idempotent semigroups. Regular elements in an ordered semigroup are studied in [34]. See also [25, 35, 36]. In this section we present a brief survey on ordered semigroups. The main references for this section are [34].

Definition 1.10.1. An ordered semigroup is a triple $(S, \cdot, \leq)$, where $(S, \cdot)$ is a semigroup, $\leq$ is a partial order on S and the following holds:

$$\forall a, b, c \in S: \; a \leq b \Rightarrow a \cdot c \leq b \cdot c \text{ and } c \cdot a \leq c \cdot b.$$

Example 18. Suppose that $S = \{a, b, c, d, e\}$. Then, $(S, \cdot, \leq)$ is an ordered semigroup, where the multiplication and order relation are defined by

(1)

$\cdot$	a	b	c	d	e
a	a	e	c	d	e
b	a	b	c	d	e
c	a	e	c	d	e
d	a	e	c	d	e
e	a	e	c	d	e

$$\leq := \{(a, a), \; (b, b), \; (c, c), \; (d, d), \; (e, e), \; (a, d), \; (c, e)\}.$$

(2)

$\cdot$	a	b	c	d	e
a	a	a	c	a	c
b	a	a	c	a	c
c	a	a	c	a	c
d	d	d	e	d	e
e	d	d	e	d	e

$$\leq := \{(a, a), \; (a, b), \; (a, c), (a, d), \; (a, e), \; (b, b), \; (b, c), \; (b, d), \; (b, e), \; (c, c), \; (c, e), \; (d, d), \; (e, e)\}.$$

(3)

$\cdot$	a	b	c	d	e
a	a	b	a	a	a
b	a	b	a	a	a
c	a	b	a	a	a
d	a	b	a	a	a
e	a	b	a	a	e

$$\leq := \{(a, a), \; (a, b), \; (a, e), \; (b, b), \; (c, b), \; (c, c), \; (c, e), \; (d, d), \; (d, b), \; (d, e), \; (e, e)\}.$$

(4)

$\cdot$	a	b	c	d	e
a	b	d	a	b	e
b	d	b	b	d	e
c	d	b	c	d	e
d	b	d	d	b	e
e	e	e	e	e	e

$$\leq := \{(a,a),\ (b,b),\ (b,c),\ (b,e),\ (c,c),\ (d,d),\ (d,a),\ (d,e),\ (e,e)\}.$$

Example 19. Let S be the system consisting of eight elements ordered by $s < q < e < t < v < f < p < u$ and with the following multiplication table:

$\cdot$	s	q	e	t	v	f	p	u
s	s	s	s	s	s	s	s	s
q	s	s	s	s	s	q	e	t
e	s	q	e	t	t	t	t	t
t	t	t	t	t	t	t	t	t
v	v	v	v	v	v	v	v	v
f	v	v	v	v	v	f	p	u
p	v	f	p	u	u	u	u	u
u	u	u	u	u	u	u	u	u

Definition 1.10.2. Let $(S, \cdot, \leq)$ be an ordered semigroup and I be a non-empty subset of S. Then, I is called a right ideal of S if the following conditions hold.

(1) $I \cdot S \subseteq I$;

(2) If $a \in I$ and $b \leq a$ with $b \in S$, then $b \in I$.

Similarly, we can define a *left ideal.* If I is both a left ideal and a right ideal of S, then I is called an *ideal* of S.

Example 20. Let $S = \{a, b, c, d, e, f\}$ be an ordered semigroup defined by the multiplication and the order below:

$\cdot$	a	b	c	d	e	f
a	a	a	a	d	a	a
b	a	b	b	d	b	b
c	a	b	c	d	e	e
d	a	a	d	d	d	d
e	a	b	c	d	e	e
f	a	b	c	d	e	f

$$\leq := \{(a,a),\ (b,b),\ (c,c),\ (d,d),\ (e,e),\ (f,e),\ (f,f)\}.$$

Then, the right ideals of S are $\{a, d\}$, $\{a, b, d\}$, $\{a, b, c, d, e\}$, and S. The left ideals of S are $\{a\}$, $\{d\}$, $\{a, b\}$, $\{a, d\}$, $\{a, b, d\}$, $\{a, b, c, d\}$, $\{a, b, d, e\}$, $\{a, b, d, e, f\}$, and S.

If two elements x and y of S generate the same principal left ideal, then we write $x \equiv y(L)$, while if x and y generate the same principal right ideal, then we write $x \equiv y(R)$. We write $x \equiv y(D)$ if there exists an element z of S such that $x \equiv z(L)$ and $z \equiv y(R)$. As is well-known, these relations are Green relations. An element x of S is called *regular* if there exists an element y of S such that

$$x \cdot y \cdot x = x \quad \text{and} \quad y \cdot x \cdot y = y. \tag{1.3}$$

When a pair (x, y) of elements of S satisfy Eq. (1.3), (x, y) is called a *regular pair* and y is called a *regular conjugate* of x. As is easily seen by Eq. (1.3), for every regular pair (x, y), both $x \cdot y$ and $y \cdot x$ are idempotents. An element x of S is called *positive* if $x \leq x^2$, while x is called *negative* if $x^2 < x$. For an element x of S, the number of distinct natural powers of x is called the *order* of x. If x is an element of finite order n, then n is the minimal natural number such that $x^n = x^{n+1}$. Evidently x is of order 1 if and only if x is idempotent. The set of all idempotents of S is denoted by $I(S)$. For an ordered semigroup $(S, \cdot, \leq)$, we call the *multiplicative dual* or, simply, *dual* of S, the ordered semigroup constructed from S by interchanging the order of multiplication but by preserving the order of S. An element z of S is said to lie between x and y, if either $x \leq z \leq y$ or $y \leq z \leq x$, while z is said to lie between x and y in the strict sense, if either $x < z < y$ or $y < z < x$.

Lemma 1.10.3. *If x and y are non-negative, then, $x \cdot y$ is non-negative. If x and y are non-positive, then $x \cdot y$ is non-positive.*

Proof. For non-negative x and y, if $x \leq y$, then $x \cdot y \leq x^3 \cdot y \leq (x \cdot y)^2$ and, if $y \leq x$, then $x \cdot y \leq x \cdot y^3 \leq (x \cdot y)^2$. The second assertion can be proved similarly. □

Corollary 1.10.4. *The set $I(S)$ of all idempotents of an ordered semigroup S, if it is non-void, is a subsemigroup of S.*

Lemma 1.10.5. *If x is non-negative, y is non-positive and $x \leq y$, then both $x \cdot y$ and $y \cdot x$ are idempotents which lie between x and y.*

Proof. We have

$$x \cdot y \leq x^3 \cdot y \leq (x \cdot y)^2 \leq x \cdot y^3 \leq x \cdot y,$$

and so $x \cdot y$ is idempotent. Moreover, we have $x \leq x^2 \leq x \cdot y \leq y^2 \leq y$. □

With respect to the order in S, the subsemigroup $I(S)$ is clearly an ordered semigroup, which plays an important role in the following discussion. As is easily seen, for $g, h \in I(S)$,

$$g \equiv h(L) \ \text{ in } S \ \Leftrightarrow \ g \cdot h = g \text{ and } h \cdot g = h, \tag{1.4}$$

$$g \equiv h(R) \ \text{ in } S \ \Leftrightarrow \ g \cdot h = h \text{ and } h \cdot g = g. \tag{1.5}$$

Hence, we obtain

$$g \equiv h(L) \text{ and } g \equiv h(R) \ \text{ in } S \ \Leftrightarrow g = h. \tag{1.6}$$

By Eqs. (1.4) and (1.5), for elements of $I(S)$, L-equivalence and R-equivalence in $I(S)$ coincide with L-equivalence and R-equivalence in S, respectively. However, for D-equivalence, such a situation does not occur. Of course, for $g, h \in I(S)$, $g \equiv h(D)$ in $I(S)$ implies $g \equiv h(D)$ in S. The D-equivalence in $I(S)$ is denoted by $D_{I(S)}$-equivalence.

Lemma 1.10.6. *If $g, h \in I(S)$ and $g \leq h$, then the following conditions are equivalent to each other:*

(1) $g \cdot h \leq h \cdot g$;
(2) $g \cdot h \cdot g = g \cdot h$;
(3) $h \cdot g \cdot h = h \cdot g$;
(4) $g \cdot h \equiv h \cdot g(L)$.

Proof. ($1 \Rightarrow 2$): We have $g \cdot h = g \cdot (g \cdot h) \leq g \cdot h \cdot g \leq (g \cdot h) \cdot h = g \cdot h$.

($2 \Rightarrow 3$): We have $h \cdot g = (h \cdot g) \cdot (h \cdot g) = h \cdot g \cdot h$.

($3 \Rightarrow 2$): It is similar.

($3 \Rightarrow 4$): If (3) holds, then both (2) and (3) hold, and so we obtain (4).

($4 \Rightarrow 1$): We have $g \cdot h = (g \cdot h) \cdot (h \cdot g) = g \cdot h \cdot g$, and so by Lemma 1.10.5, $g \leq g \cdot h = g \cdot h \cdot g \leq h \cdot g \leq h$. □

Lemma 1.10.7. *If $g, h \in I(S)$ and $g \leq h$, then the following conditions are equivalent to each other:*

(1) $h \cdot g \leq g \cdot h$;
(2) $h \cdot g \cdot h = g \cdot h$;
(3) $g \cdot h \cdot g = h \cdot g$;
(4) $g \cdot h \equiv h \cdot g(R)$.

Proof. It is similar to the proof of Lemma 1.10.6. □

Corollary 1.10.8. *We have $g \cdot h \equiv h \cdot g(D_{I(S)})$, for all $g, h \in I(S)$.*

Lemma 1.10.9. *In an ordered idempotent semigroup, each D-class consists of either only one L-class or only one R-class.*

Proof. Let L_1 and L_2 be two L-classes in a D-class D, and let R_1 and R_2 be two R-classes in D. We denote the intersection elements of L_1 and R_1, of L_1 and R_2, of L_2 and R_1, and of L_2 and R_2 by a, b, c, and d, respectively. Without loss of generality, we assume that $d \leq a$. Then, we have $d = b \cdot d \leq b \cdot a = b = d \cdot b \leq a \cdot b = a$. If $b \leq c$, then we have, in a similar way, $b \leq d \leq c$, and so $b = d$, from which it follows that

$L_1 = L_2$. On the other hand, if $c \leq b$, then we obtain $R_1 = R_2$ similarly. This completes the proof. □

Corollary 1.10.10. *Each $D_{I(S)}$-equivalence class in $I(S)$ consists of either only one L-equivalence class in $I(S)$ or only one R-equivalence class in $I(S)$.*

A regular pair (x, y) is said to be of order n, if both x and y are elements of order n. A regular pair of order 1 is also called an *idempotent regular pair.* Now, we give a theorem which characterizes idempotent regular pairs.

Theorem 1.10.11

(1) *For a regular pair (x, y) of S, x is idempotent if and only if y is idempotent.*

(2) *For $g, h \in I(S)$, (g, h) is a regular pair if and only if $g \equiv h(D_{I(S)})$.*

Proof. (1) Suppose that (x, y) is a regular pair and that x is an idempotent. Then, $y = y \cdot x \cdot y = (y \cdot x) \cdot (x \cdot y)$ is an idempotent, by Corollary 1.10.4.

(2) First, suppose that (g, h) is an idempotent regular pair and that $g \leq h$. By Eq. (1.3),

$$g \equiv h \cdot g(L), \quad h \equiv g \cdot h(L), \quad g \equiv g \cdot h(R), \quad h \equiv h \cdot g(R).$$

If $g \cdot h \leq h \cdot g$, then, by Lemma 1.10.6, $h \cdot g \equiv g \cdot h(L)$, and so $g \equiv h(L)$. If $h \cdot g \leq g \cdot h$, then we obtain $g \equiv h(R)$ similarly. Next, suppose that $g \equiv h(D_{I(S)})$. Then, by Corollary 1.10.10, either $g \equiv h(L)$ or $g \equiv h(R)$. If $g \equiv h(L)$, then $g \cdot h \cdot g = (g \cdot h) \cdot g = g^2 = g$, $h \cdot g \cdot h = (h \cdot g) \cdot h = h^2 = h$, and so (g, h) is a regular pair. In the case when $g \equiv h(R)$, we obtain the same result in a similar way. □

Lemma 1.10.12. *For two regular pairs (x, y) and (z, w), $(x \cdot z, w \cdot y)$ is a regular pair.*

Proof. By Corollary 1.10.4, both $y \cdot x \cdot z \cdot w$ and $z \cdot w \cdot y \cdot x$ are idempotent. Hence, we obtain

$$\begin{aligned}(x \cdot z) \cdot (w \cdot y) \cdot (x \cdot z) &= (x \cdot y \cdot x) \cdot z \cdot w \cdot y \cdot x \cdot (z \cdot w \cdot z) \\ &= x \cdot (y \cdot x \cdot z \cdot w) \cdot z = x \cdot z,\end{aligned}$$

$$\begin{aligned}(w \cdot y) \cdot (x \cdot z) \cdot (w \cdot y) &= (w \cdot z \cdot w) \cdot y \cdot x \cdot z \cdot w \cdot (y \cdot x \cdot y) \\ &= w \cdot (z \cdot w \cdot y \cdot x) \cdot y = w \cdot y.\end{aligned}$$

□

Corollary 1.10.13. *If (x, y) is a regular pair, then, for every natural number n, (x^n, y^n) is a regular pair.*

Corollary 1.10.14. *The set of all regular elements of S, if it is non-void, is a subsemigroup of S.*

Lemma 1.10.15. *If (p, q) is a regular pair such that $q \leq p$, then q is non-positive and p is non-negative.*

Proof. We have $q^3 \leq q \cdot p \cdot q = q$ and $p = p \cdot q \cdot p \leq p^3$, from which the lemma follows immediately. □

Lemma 1.10.16. *Let (p, q) be a regular pair such that $q \leq p$ and $q{\cdot}p \leq p{\cdot}q$. Then, the following six conditions are equivalent to each other:*

(1) $p \cdot q^2 = p \cdot q^2 \cdot p$;

(2) $q^2 \cdot p = q^2$;

(3) $q^2 = q^3$;

(4) $q \cdot p^2 = q \cdot p^2 \cdot q$;

(5) $p^2 \cdot q = p^2$;

(6) $p^2 = p^3$.

Moreover, these conditions imply

(7) $(q \cdot p) \cdot (p \cdot q) \equiv q^2 \equiv p^2 \equiv (p \cdot q) \cdot (q \cdot p)(L)$.

Proof. (1 ⇒ 2): We have $q^2 \cdot p = q \cdot p \cdot q^2 \cdot p = q \cdot p \cdot q^2 = q^2$.

(2 ⇒ 1): We have $q^2 \cdot p^2 = q^2 \cdot p = q^2$ and so q^2 is an idempotent, by Corollary 1.10.13.

(3 ⇒ 1): We have

$$p \cdot q^2 = p \cdot q^3 \leq p \cdot q^2 \cdot p = p \cdot q^2 \cdot (q \cdot p) \leq p \cdot q^2 \cdot (p \cdot q) = p \cdot q^2.$$

Similarly the conditions (4)–(6) are equivalent to each other.

(3 ⇒ 6): By Theorem 1.10.11 and Corollary 1.10.13, (p^2, q^2) is an idempotent regular pair.

(6 ⇒ 3): It is similar.

This proves the first half of the lemma. Next suppose that these conditions hold. Then, we obtain

$$(q \cdot p^2 \cdot q) \cdot q^2 = ((q \cdot p^2 \cdot q) \cdot q) \cdot q = q \cdot p^2 \cdot q,$$

$$q^2 \cdot (q \cdot p^2 \cdot q) = ((q^3 \cdot p) \cdot p) \cdot q = q^2,$$

$$q^2 \cdot p^2 = (q^2 \cdot p) \cdot p = q^2,$$

$$p^2 \cdot q^2 = (p^2 \cdot q) \cdot q = p^2,$$

$$p^2 \cdot (p \cdot q^2 \cdot p) = ((p^3 \cdot q) \cdot q) \cdot p = p^2,$$

$$(p \cdot q^2 \cdot p) \cdot p^2 = ((p \cdot q^2 \cdot p) \cdot p) \cdot p = p \cdot q^2 \cdot p.$$

Hence, (7) holds. □

Lemma 1.10.17. *Let (p, q) be a regular pair such that $q \leq p$ and $p \cdot q \leq q \cdot p$. Then, the following six conditions are equivalent to each other:*

(1) $q^2 \cdot p = p \cdot q^2 \cdot p$;
(2) $p \cdot q^2 = q^2$;
(3) $q^2 = q^3$;
(4) $p^2 \cdot q = q \cdot p^2 \cdot q$;
(5) $q \cdot p^2 = p^3$;
(6) $p^2 = p^3$.

Moreover, these conditions imply

(7) $(q \cdot p) \cdot (p \cdot q) \equiv q^2 \equiv p^2 \equiv (p \cdot q) \cdot (q \cdot p)(R)$.

Proof. It is similar to the proof of Lemma 1.10.16. □

Corollary 1.10.18. *For a regular pair (x, y), x is an element of order 2 if and only if y is of order 2.*

Theorem 1.10.19. *If (x, y) is a regular pair such that either x or y is an element of finite order, then (x, y) is a regular pair of order either 2 or 1.*

Proof. By Theorem 1.10.11 and Corollary 1.10.18, it suffices to show that if x is an element of finite order, then x is of order at most 2. Here we prove this assertion only in the case when $x \leq y$ and $x \cdot y \leq y \cdot x$. Then, by Lemma 1.10.15, x is non-positive. Now, suppose it were true that $x^{n+1} = x^n < x^{n-1}$ for a natural number $n \geq 3$. Then, $x \cdot y \cdot x^n = x^n < x^{n-1} = x \cdot x^{n-2}$ and so $y \cdot x^n < x^{n-2}$. Hence, $y \cdot x^{n+1} \leq x^{n-1}$. On the other hand,

$$x^{n-1} = x \cdot y \cdot x^{n-1} \leq (y \cdot x) \cdot x^{n-1} = y \cdot x^n = y \cdot x^{n+1}.$$

Hence, $x^{n-1} = y \cdot x^{n+1}$. Then, we would have

$$x^n = x^{n-1} \cdot x = y \cdot x^{n+2} = y \cdot x^{n+1} = x^{n-1},$$

which is a contradiction. □

CHAPTER 2

Semihypergroups

2.1 HISTORY OF ALGEBRAIC HYPERSTRUCTURES

The algebraic hyperstructure notion was introduced in 1934 by Marty [37], at the 8th Congress of Scandinavian Mathematicians. He published some notes on hypergroups, using them in different contexts: algebraic functions, rational fractions, non-commutative groups. Hypergroups are a suitable generalization of groups. We know in a group, the composition of two elements is an element, while in a hypergroup, the composition of two elements is a set.

In [38], Prenowitz represented several kinds of geometries (projective, descriptive, and spherical) as hypergroups, and later, with Jantosciak [39], founded geometries on join spaces, a special kind of hypergroups, which in the last decades were shown to be useful instruments in the study of several matters: graphs, hypergraphs, and binary relations.

Several kinds of hypergroups have been intensively studied, such as regular hypergroups, reversible regular hypergroups, canonical hypergroups, cogroups, and cyclic hypergroups. The situations that occur in hypergroup theory, are often extremely diversified and complex with respect to group theory. For instance, there are homomorphisms of various types between hypergroups and there are several kinds of subhypergroups, such as closed, invertible, ultraclosed, and conjugable.

Around the 1940s, the general aspects of the theory, the connections with groups and various applications in geometry were studied in France by F. Marty, M. Krasner, M. Kuntzmann, and R. Croisot; in USA by M. Dresher, O. Ore, W. Prenowitz, H.S. Wall, J.E. Eaton, H. Campaigne, and L. Griffiths; in Russia by A. Dietzman and A. Vikhrov; in Italy by G. Zappa; and in Japan by Y. Utumi.

Over the following two decades, other interesting results on hyperstructures were obtained; for instance, in Italy, A. Orsatti studied semiregular hypergroups; in Czechoslovakia, K. Drbohlav studied hypergroups of two-sided classes; and in Romania, M. Benado studied hyperlattices.

Semihypergroup Theory
http://dx.doi.org/10.1016/B978-0-12-809815-8.00002-4

The theory witnessed an important progress starting with the 1970s, when its research area enlarged. In France, M. Krasner, M. Koskas, and Y. Sureau investigated the theory of subhypergroups and the relations defined on hyperstructures; in Greece, J. Mittas and his students M. Konstantinidou, K. Serafimidis, S. Ioulidis, and C.N. Yatras studied the canonical hypergroups, the hyperrings, and the hyperlattices; Ch. Massouros obtained important results about hyperfields and other hyperstructures. G. Massouros, together with J. Mittas, studied applications of hyperstructures to Automata. D. Stratigopoulos continued some of Krasner's ideas, studying in depth non-commutative hyperrings and hypermodules. T. Vougiouklis, L. Konguetsof, and later S. Spartalis and A. Dramalidis analyzed especially the cyclic hypergroups, the *P*-hyperstructures.

Significant contributions to the study of regular hypergroups, complete hypergroups, of the heart and of the hypergroup homomorphisms in general or with applications in Combinatorics and Geometry were brought by the Italian mathematician P. Corsini and his research group, among whom we mention M. de Salvo, R. Migliorato, F. de Maria, G. Romeo, and P. Bonansinga.

Also around 1970s, some connections between hyperstructures and ordered systems, particularly lattices, were established by T. Nakano and J.C. Varlet. Around the 1980s and 1990s, associativity semihypergroups were analyzed in the context of semigroup theory by T. Kepka and then by J. Jezec, P. Nemec, and K. Drbohlav; and in Finland by M. Niemenmaa.

In USA, R. Roth used canonical hypergroups in solving some problems of character theory of finite groups, while S. Comer studied the connections among hypergroups, combinatorics, and the relation theory. J. Jantosciak continued the study of join spaces, introduced by W. Prenowitz; he considered a generalization of them for the noncommutative case and studied correspondences between homomorphisms and the associated relations.

In North America, hyperstructures have been studied both in the USA (at Charleston, South Carolina—The Citadel, New York-Brooklyn College, CUNY, Cleveland, Ohio—John Carroll University) and in Canada (at Université de Montréal).

A big role in spreading this theory is played by the Congresses on Algebraic Hyperstructures and their Applications.

The first three Congresses were organized by P. Corsini in Italy. The contribution of P. Corsini to the development of Hyperstructure Theory has been decisive. He has delivered lectures about hyperstructures and their applications in several countries, several times, for instance, in Romania, Thailand, Iran, China, and Montenegro, making known this theory. After his visits in these countries, hyperstructures have had a substantial development.

Coming back to the Congresses on Algebraic Hyperstructures, the first two were organized in Taormina, Sicily, in 1978 and 1983, with the names: "Sistemi Binari e loro Applicazioni" and "Ipergruppi, Strutture Multivoche e Algebrizzazione di Strutture d'Incidenza." The third Congress, called "Ipergruppi, altre Strutture Multivoche e loro Applicazioni," was organized in Udine in 1985.

The fourth Congress, organized by T. Vougiouklis in Xanthi in 1990, used the name of "Algebraic Congress on Hyperstructures and their Applications", also known as AHA Congress. Since 1990, AHA Congresses have been organized every three years. From the 1990s Hyperstructure Theory has been a constant concern also for the Romanian mathematicians, the decisive moment being the fifth AHA Congress, organized in 1993 at the University "Al.I.Cuza" of Iasi by M. Stefanescu. This domain of modern algebra is a topic of great interest also for Romanian researchers, who have published a lot of papers on hyperstructures in national or international journals, have given communications in conferences and congresses, and have written Ph.D. theses in this field.

The sixth AHA Congress was organized in 1996 at the Agriculture University of Prague by T. Kepka and P. Nemec; the seventh was organized in 1999 by R. Migliorato in Taormina, Sicily; then the eighth was organized in 2002 by T. Vougiouklis in Samothraki, Greece. All these congresses were organized in Europe. Nowadays, one works successfully on Hyperstructures in the following countries of Europe:

- in Greece, at Thessaloniki (Aristotle University), at Alexandropoulis (Democritus University of Thrace), at Patras (Patras University), Orestiada (Democritus University of Thrace), and at Athens;
- in Italy, at Udine University, at Messina University, at Rome (Universita' "La Sapienza"), at Pescara (D'Annunzio University), at Teramo (Universita' di Teramo), and at Palermo University;
- in Romania, at Iasi ("Al.I. Cuza" University), Cluj ("Babes-Bolyai" University), and Constanta ("Ovidius" University);
- in the Czech Republic, at Praha (Charles University, Agriculture University), at Brno (Brno University of Technology, Military Academy of Brno, Masaryk University), and Olomouc (Palacky University);
- in Montenegro, at Podgorica University.

Let us continue with the following AHA Congresses.

The ninth congress on hyperstructures, organized in 2005 by R. Ameri in Babolsar, Iran, was the first of this kind in Asia. In the past millennia, Iran gave fundamental contributions to mathematics, and in particular, to algebra (for instance, Khwarizmi, Kashi, Khayyam, and recently Zadeh). Many scientists have well understood the importance of hyperstructures, on the theoretical point of view, and for the applications to a wide variety of scientific sectors.

Nowadays, hyperstructures are cultivated in many universities and research centers in Iran, among which we mention Yazd University, Shahid Bahonar University of Kerman, Mazandaran University, Kashan University, Ferdowsi University of Mashhad, Tehran University, Tarbiat Modarres University, Zahedan (Sistan and Baluchestan University), Semnan University, Islamic Azad University of Kerman, Shahid Beheshti University of Tehran, and the Center for Theoretical Physics and Mathematics of Tehran, Zanjan (Institute for Advanced Studies in Basic Sciences).

Another Asian country where hyperstructures have had success is Thailand. In Chulalornkorn University of Bangkok, important results have been obtained by Y. Kemprasit and her students Y. Punkla, S. Chaoprakhoi, N. Triphop, and C. Namnak on the connections among hyperstructures, semigroups, and rings. In Khon Kaen University, T. Changphas and B. Pibaljommee have obtained important results on the connection of ordered semihypergroups.

There are other Asian centers for researches in hyperstructures. We mention here India (University of Calcutta; Aditanar College of Arts and Sciences, Tiruchendur, Tamil Nadu); Korea (Chiungju National University, Chiungju National University of Education, Gyeongsang National University, Jinju); Japan (Hitotsubashi University of Tokyo); the Sultanate of Oman (Education College for Teachers); and China (Northwest University of Xian, Yunnan University of Kunming).

Hyperstructures have been also cultivated in Germany, the Netherlands, Belgium, Macedonia, Serbia, Slovakia, Spain, Uzbekistan, and Australia. The tenth AHA Congress was held in Brno, Czech Republic in the autumn of 2008. It was organized by Šárka Hošková, at the Military Academy of Brno. The eleventh AHA Congress was held in Pescara, Italy in the autumn of 2011. It was organized by A. Maturo, at the Università degli Studi "G. d'Annunzio" Chieti-Pescara. The twelfth AHA Congress was held in Xanthi, Greece in September 2014. It was organized by S. Spartalis at the Democritus University of Thrace.

More than 850 papers and some books have been written on hyperstructures. Many of them are dedicated to the applications of hyperstructures in other topics. We shall mention here some of the fields connected with hyperstructures and only some names of mathematicians who have worked in each topic:

- *Biology* (B. Davvaz, A. Dehghan Nezad, M. M. Heidari, M. Ghadiri, R. Nekouian),
- *Physics* (R. Santilli, T. Vougiouklis, B. Davvaz, S. Hošková, J. Chvalina, P. Račková, A. Dehghan Nezhad, S.M. Moosavi Nejad, M. Nadjafikhah),
- *Chemistry* (B. Davvaz, A. Dehghan Nezhad, A. Benvidi, M. Mazloum-Ardakani, S.-C. Chung, K. M. Chun, N.J. Kim, S.Y. Jeong, H. Sim, J. Lee, H. Maeng),
- *Geometry* (W. Prenowitz, J. Jantosciak, and later G. Tallini),
- *Codes* (G. Tallini, B. Davvaz, T. Musavi),
- *Cryptography and Probability* (L. Berardi, F. Eugeni, S. Innamorati, A. Maturo),
- *Automata* (G. Massouros, J. Chvalina, L. Chvalinova),
- *Artificial Intelligence* (G. Ligozat),
- *Median Algebras, Relation Algebras, C-algebras* (S. Comer),
- *Boolean Algebras* (A.R. Ashrafi, M. Konstantinidou, B. Davvaz),
- *Categories* (M. Scafati, M.M. Zahedi, C. Pelea, R. Bayon, N. Ligeros, S.N. Hosseini, B. Davvaz, M. Alp, M.R. Khosharadi-Zadeh),
- *Topology* (J. Mittas, M. Konstantinidou, M.M. Zahedi, R. Ameri, S. Hošková, B. Davvaz, D. Heidari, S. M. S. Modarres),
- *Binary Relations* (J. Chvalina, I.G. Rosenberg, P. Corsini, V. Leoreanu, D. Freni, B. Davvaz, S. Spartalis, I. Chajda, S. Hošková, I. Cristea, M. De Salvo, G. Lo Faro, S.M. Anvariyeh, S. Mirvakili, M. Jafarpour, H. Aghabozorgi),
- *Graphs and Hypergraphs* (P. Corsini, I.G. Rosenberg, V. Leoreanu, M. Gionfriddo, A. Iranmanesh, M.R. Khosharadi-Zadeh, B. Davvaz, M. Farshi, S. Mirvakili),
- *Lattices and Hyperlattices* (J.C. Varlet, T. Nakano, J. Mittas, A. Kehagias, M. Konstantinidou, K. Serafimidis, V. Leoreanu, I.G. Rosenberg, B. Davvaz, S. Rasouli, G. Calugareanu, G. Radu, A.R. Ashrafi),
- *Fuzzy Sets and Rough Sets* (P. Corsini, M.M. Zahedi, B. Davvaz, R. Ameri, R.A. Borzooei, V. Leoreanu, I. Cristea, A. Kehagias, A. Hasankhani, I. Tofan, C. Volf, G.A. Moghani, H. Hedayati),
- *Intuitionistic Fuzzy Hyperalgebras* (B. Davvaz, E. Hassani Sadrabadi, I. Cristea, R.A. Borzooei, Y.B. Jun, W.A. Dudek, L. Torkzadeh),
- *Generalized Dynamical Systems* (M.R. Molaei, B. Davvaz, A. Dehghan Nezad).

Another topic that has aroused the interest of several mathematicians, is that of H_v-structures, introduced by T. Vougiouklis and then studied also by B. Davvaz, M.R. Darafsheh, M. Ghadiri, R. Migliorato, S. Spartalis, A. Dramalidis, A. Iranmanesh, M.N. Iradmusa, and A. Madanshekaf. H_v-structures are a special kind of hyperstructures, for which the weak associativity holds. Recently, n-ary hyperstructures, introduced by B. Davvaz and T. Vougiouklis, represent an intensively studied field of research.

Therefore, there are good reasons to hope that Hyperstructure Theory will be one of the more successful fields of research in algebra.

2.2 SEMIHYPERGROUP AND EXAMPLES

The concept of a semihypergroup is a generalization of the concept of a semigroup. Many authors studied different aspects of semihypergroups, for instance, Anvariyeh et al. [40], Bonansinga and Corsini [41], Corsini et al. [42–44], Davvaz [45–47], Davvaz and Leoreanu-Fotea [48], Davvaz and Poursalavati [49], De Salvo et al. [50–54], Fasino and Freni [55, 56], Freni [57], Guan [58], Hila et al. [59–61], Kudryavtseva and Mazorchuk [62], Leoreanu [63], Mousavi et al. [64–66], Onipchuk [67], Savettaseranee et al. [68], and many others.

A *hypergroupoid* $(H, \circ)$ is a non-empty set H together with a map $\circ : H \times H \to \mathcal{P}^*(H)$ called *(binary) hyperoperation*, where $\mathcal{P}^*(H)$ denotes the set of all non-empty subsets of H. The image of the pair (x, y) is denoted by $x \circ y$.

If A, B are non-empty subsets of H and $x \in H$, then by $A \circ B$, $A \circ x$, and $x \circ B$ we mean

$$A \circ B = \bigcup_{\substack{a \in A \\ b \in B}} a \circ b, \quad A \circ x = A \circ \{x\} \text{ and } x \circ B = \{x\} \circ B.$$

Definition 2.2.1. A hypergroupoid $(H, \circ)$ is called a *semihypergroup* if

$$(x \circ y) \circ z = x \circ (y \circ z),$$

for all $x, y, z \in H$. This means that

$$\bigcup_{u \in x \circ y} u \circ z = \bigcup_{v \in y \circ z} x \circ v.$$

A semihypergroup H is *finite* if it has only a finitely many elements. A semihypergroup H is *commutative* if it satisfies

$$x \circ y = y \circ x,$$

for all $x, y \in H$.

Remark 7. Every semigroup is a semihypergroup.

Remark 8. The associativity for semihypergroups can be applied for subsets, ie, if $(H, \circ)$ is a semihypergroup, then for all non-empty subsets A, B, C of H, we have $(A \circ B) \circ C = A \circ (B \circ C)$.

The element $a \in H$ is called *scalar* if

$$|a \circ x| = |x \circ a| = 1,$$

for all $x \in H$. An element e in a semihypergroup $(H, \circ)$ is called *scalar identity* if

$$x \circ e = e \circ x = \{x\},$$

for all $x \in H$. An element e in a semihypergroup $(H, \circ)$ is called *identity* if

$$x \in e \circ x \cap x \circ e,$$

for all $x \in H$. An element $a' \in H$ is called an *inverse* of $a \in H$ if there exists an identity $e \in H$ such that

$$e \in a \circ a' \cap a' \circ a.$$

An element 0 in a semihypergroup $(H, \circ)$ is called *zero element* if $x \circ 0 = 0 \circ x = \{0\}$, for all $x \in H$.

Similar to semigroups, we can describe the hyperoperation on a semihypergroup by Cayley table.

Example 21

(1) Let $H = \{a, b, c, d\}$. Define the hyperoperation $\circ$ on H by the following table.

$\circ$	a	b	c	d
a	a	$\{a, b\}$	$\{a, c\}$	$\{a, d\}$
b	a	$\{a, b\}$	$\{a, c\}$	$\{a, d\}$
c	a	b	c	d
d	a	b	c	d

Then, $(H, \circ)$ is a semihypergroup.

(2) Let $H = \{a, b, c, d, e\}$. Define the hyperoperation $\circ$ on H by the following table.

$\circ$	a	b	c	d	e
a	a	$\{a, b, d\}$	a	$\{a, b, d\}$	$\{a, b, d\}$
b	a	b	a	$\{a, b, d\}$	$\{a, b, d\}$
c	a	$\{a, b, d\}$	$\{a, c\}$	$\{a, b, d\}$	$\{a, b, c, d, e\}$
d	a	$\{a, b, d\}$	a	$\{a, b, d\}$	$\{a, b, d\}$
e	a	$\{a, b, d\}$	$\{a, c\}$	$\{a, b, d\}$	$\{a, b, c, d, e\}$

Then, $(H, \circ)$ is a semihypergroup.

(3) Let H be the unit interval $[0, 1]$. For every $x, y \in H$, we define

$$x \circ y = \left[0, \frac{xy}{2}\right].$$

Then, $(H, \circ)$ is a semihypergroup.

(4) Let $\mathbb{N}$ be the set of non-negative integers. We define the following hyperoperation on $\mathbb{N}$,

$$x \circ y = \{z \in \mathbb{N} \mid z \geq \max\{x, y\}\},$$

for all $x, y \in \mathbb{N}$. Then, $(\mathbb{N}, \circ)$ is a semihypergroup.

(5) Let $(S, \cdot)$ be a semigroup and K be any subsemigroup of S. Then, the set $S/K = \{x \cdot K \mid x \in S\}$ becomes a semihypergroup, where the hyperoperation is defined in a usual manner $\overline{x} \circ \overline{y} = \{\overline{z} \mid z \in \overline{x} \cdot \overline{y}\}$ with $\overline{x} = x \cdot K$.

(6) The set of real numbers $\mathbb{R}$ with the following hyperoperation

$$a \circ b = \begin{cases} (a, b) & \text{if } a < b \\ (b, a) & \text{if } b < a \\ \{a\} & \text{if } a = b, \end{cases}$$

for all $a, b \in \mathbb{R}$ is a semihypergroup, where (a, b) is the open interval $\{x \mid a < x < b\}$.

(7) Let $(S, \cdot)$ be a semigroup and P a non-empty subset of S. We define the following hyperoperation on S,

$$x \circ_P y = x \cdot P \cdot y,$$

for all $x, y \in S$. Then, $(S, \circ_P)$ is a semihypergroup. The hyperoperation $\circ_P$ is called *P-hyperoperation*.

(8) Let $(S, \cdot)$ be a semigroup and for all $x, y \in S$, $\langle x, y \rangle$ denotes the subsemigroup generated by x and y. We define $x \circ y = \langle x, y \rangle$. Then, $(S, \circ)$ is a semihypergroup.

(9) Let $(H, \circ)$ and $(H', \star)$ be two semihypergroups. Then, the Cartesian product of these two semihypergroups is a semihypergroup with the following hyperoperation

$$(x, y) \otimes (x', y') = \{(a, b) \mid a \in x \circ x', \ b \in y \star y'\},$$

for all $(x, y), (x', y') \in H \times H'$.

(10) If $(H, \circ)$ is a semihypergroup, then H is also a semihypergroup with respect to the hyperoperation $\star$ defined by $x \star y = y \circ x$, for all $x, y \in H$. The semihypergroup $(H, \star)$ is called the *transposed* of $(H, \circ)$ and is denoted by H^T. Clearly, the use of this term is motivated by the fact that, in the finite case, the multiplication table of H^T is obtained by transposing the multiplication table of $(H, \circ)$.

Definition 2.2.2. A semihypergroup $(H, \circ)$ is called a *hypergroup* if

$$a \circ H = H \circ a = H,$$

for all $a \in H$. The above conditions are called the *reproduction axioms*.

A hypergroup is called *regular* if it has at least one identity and each element has at least one inverse.

One can find many examples and a deep discussion about hypergroups in [23, 39, 69–72].

Remark 9. One of the generalizations of semihypergroups is Γ-semihypergroups. This concept is investigated in [73–81]. Another generalization of semihypergroups is H_v-semigroups. For the results on this concept, see [17, 82–84].

2.3 REGULAR SEMIHYPERGROUPS

Similar to semigroups, one can define the notion of regular semihypergroups. Chaopraknoi and Triphop [85], Asokkumar and Velrajan [86], and Jafarabadi et al. [87] studied regular semihypergroups. In this section, we investigate regular semihypergroups and their properties. The main reference for this section is [86].

Definition 2.3.1. An element x of a semihypergroup $(H, \circ)$ is said to be *regular* if $x \in x \circ H \circ x$. That is, there exists an element $y \in H$ such that $x \in x \circ y \circ x$. Now, the element y is called a *generalized inverse* of x. Usually generalized inverses of x are not unique. A semihypergroup $(H, \circ)$ is said to be *regular* if every element of H is regular.

Remark 10. Every regular semigroup is a regular semihypergroup.

Proposition 2.3.2. *Any hypergroup $(H, \circ)$ is a regular semihypergroup.*

Proof. Let x be an element of the hypergroup H. Since $H = H \circ x$, it follows that $x \in a \circ x$ for some $a \in H$. Since $a \in H = x \circ H$, it follows that $a \in x \circ y$ for some $y \in H$. Thus, $x \in a \circ x \subseteq x \circ y \circ x$. Hence, $(H, \circ)$ is regular. □

Example 22

(1) Suppose that a hyperoperation $\circ$ on $H = \{a, b, c\}$ is defined by the following table

$\circ$	a	b	c
a	a	$\{a, b\}$	$\{a, b, c\}$
b	a	$\{a, b\}$	$\{a, b, c\}$
c	a	$\{a, b\}$	$\{c\}$

Then, $(H, \circ)$ is a regular semihypergroup and it is neither commutative nor reproductive. Also, each element of H is a generalized inverse of every element of H.

(2) Let H be a non-empty set and $\{P_1, P_2, \ldots, P_n\}$ be a partition of H. For $x \in P_i$ and $y \in P_j$, define a hyperoperation $\circ$ on H by $x \circ y = P_k$, where $k = \max\{i, j\}$. Then, $(H, \circ)$ is a semihypergroup. Let $a \in H$. If $a \in P_i$ for some $1 \leq i \leq n$, then $a \circ a \circ a = P_i$ and so $a \in a \circ a \circ a$. Thus, $(H, \circ)$ is a regular semihypergroup. Also, for $a \in P_j$ for some $j > 1$,

$$a \circ H = a \circ \left(\bigcup_{i=1}^{n} P_i \right) = \bigcup_{i=1}^{n} (a \circ P_i) = \bigcup_{i=j}^{n} P_i \neq H.$$

Hence, $(H, \circ)$ is not a hypergroup.

(3) Consider Example 21(9). If $(H, \circ)$ and $(H', \star)$ are regular semihypergroups, then $H \times H'$ is a regular semihypergroup too.

Let $(S, \cdot)$ be a semigroup and P a non-empty subset of S. Define a hyperoperation $\circ_P$ on S by $x \circ_P y = x \cdot P \cdot y$, for all $x, y \in S$. Then, $(S, \circ_P)$ is a semihypergroup (Example 21(7)). We denote this semihypergroup induced by P on S by (S, P). In the continuation of this section, S denotes a semigroup and P denotes a non-empty subset of S. If $a \in S$ is regular in S, then we denote $V(a) = \{x \in S | a = a \cdot x \cdot a\}$.

Proposition 2.3.3. *The element $a \in S$ is regular in (S, P) if and only if a is regular in S and $V(a) \cap P \cdot S \cdot P \neq \emptyset$. That is, a is regular in S and there exists a generalized inverse of a of the form $p \cdot s \cdot q$ for some $p, q \in P$ and $s \in S$.*

Proof. Let a be regular in (S, P). Then, there exists an element $x \in S$ such that $a \in a \circ_P x \circ_P a = a \cdot P \cdot x \cdot P \cdot a$. That is, there exist elements

$p, q \in P$ such that $a = a \cdot p \cdot x \cdot q \cdot a = a \cdot y \cdot a$, where $y = p \cdot x \cdot q \in P \cdot S \cdot P$. Hence, $y \in V(a) \cap P \cdot S \cdot P$. Therefore, $V(a) \cap P \cdot S \cdot P \neq \emptyset$.

Conversely, suppose that a is regular in S and $V(a) \cap P \cdot S \cdot P \neq \emptyset$. Then, there exists an element $x \in V(a) \cap P \cdot S \cdot P$. That is, $a = a \cdot x \cdot a$ and $x = p \cdot s \cdot q$ for some $p, q \in P$ and $s \in S$. Hence, $a = a \cdot (p \cdot s \cdot q) \cdot a \in a \cdot P \cdot s \cdot P \cdot a = a \circ p s \circ p a$. Therefore, a is regular in (S, P). □

Corollary 2.3.4. *If $a \in S$ is regular in the semihypergroup (S, P), then the element a is regular in the semigroup S.*

Corollary 2.3.5. *If the semihypergroup (S, P) is regular, then the semigroup S is regular.*

Example 23. Define a binary operation $\star$ on the set of all natural numbers $\mathbb{N}$, by $a \star b = \max\{a, b\}$, for all $a, b \in \mathbb{N}$. Then, $(\mathbb{N}, \star)$ is a semigroup. Since $a = a \star a = a \star a \star a$, for all $a \in \mathbb{N}$, it follows that $\mathbb{N}$ is a regular semigroup. Let $c \in \mathbb{N}$. Let P be any non-empty subset of $\{c + 1, c + 2, \ldots\}$. The element c is not regular in the semihypergroup $(\mathbb{N}, P)$. This example shows that there are subsets P of $\mathbb{N}$ such that the converse of Corollary 2.3.4 need not be true. Also, we note that $V(c) = \{1, 2, 3, \ldots, c\}$ and $V(c) \cap P \star \mathbb{N} \star P = \emptyset$.

Theorem 2.3.6. *(S, P) is regular if and only if S is regular and $V(a) \cap P \cdot S \cdot P \neq \emptyset$, for every $a \in S$.*

Proof. The proof follows from Proposition 2.3.3. □

Now, we shall obtain sufficient conditions on S and P so that the converse of Corollary 2.3.4 holds.

Proposition 2.3.7. *If $a \in S$ is regular, then a is regular in (S, S).*

Proof. Since $a \in S$ is regular, there exists an element $x \in S$ such that $a = a \cdot x \cdot a$. Since $a = a \cdot x \cdot a = a \cdot (x \cdot a \cdot x) \cdot a$, it follows that $x \cdot a \cdot x \in V(a) \cap S \cdot S \cdot S$. Hence, $V(a) \cap S \cdot S \cdot S \neq \emptyset$. Therefore, a is regular in (S, S). □

Corollary 2.3.8. *If S is regular, then the semihypergroup (S, S) is regular.*

Proposition 2.3.9. *If $a \in S$ is regular, then a is regular in $(S, V(a))$.*

Proof. Since $a \in S$ is regular, there exists an element $x \in S$ such that $a = a \cdot x \cdot a$. Then, $x \in V(a)$. Since $a = a \cdot x \cdot a = a \cdot (x \cdot a \cdot x) \cdot a$, it follows that $x \cdot a \cdot x \in V(a) \cap V(a) \cdot S \cdot V(a)$. Hence, $V(a) \cap V(a) \cdot S \cdot V(a) \neq \emptyset$. Therefore, a is regular in $(S, V(a))$. □

Proposition 2.3.10. *Let P, Q be non-empty subsets of S such that $P \subseteq Q$. If a is regular in the semihypergroup (S, P), then the element a is regular in the semihypergroup (S, Q).*

Proof. Let $a \in S$. Since the semihypergroup (S, P) is regular, it follows that $V(a) \cap P \cdot S \cdot P \neq \emptyset$. Since $P \subseteq Q$, it follows that $V(a) \cap Q \cdot S \cdot Q \neq \emptyset$. That is, a is regular in the semihypergroup (S, Q). □

Corollary 2.3.11. *Let P, Q be non-empty subsets of S such that $P \subseteq Q$. If the semihypergroup (S, P) is regular, then the semihypergroup (S, Q) is regular.*

Corollary 2.3.12. *If S is regular, then $(S, \bigcup_{a \in S} V(a))$ is a regular semihypergroup.*

Proof. Let $a \in S$. Then, by Corollary 2.3.9, a is regular in $(S, V(a))$. Hence, by Proposition 2.3.10, a is regular in $(S, \bigcup_{a \in S} V(a))$. □

Proposition 2.3.13. *Let $a \in S$ be regular in S. Let P be a non-empty subset of S such that $a \in a \cdot P^m \cap P^n \cdot a$, for some natural numbers m, n. Then, the element a is regular in the semihypergroup (S, P).*

Proof. Since a is a regular element of the semigroup S, there exists an element $b \in S$ such that $a = a \cdot b \cdot a$. Since $a \in a \cdot P^m \cap P^n \cdot a$, it follows that there exists $c \in P^m$ and $d \in P^n$ such that $a = a \cdot c$ and $a = d \cdot a$. Therefore, $a = a \cdot b \cdot a = a \cdot c \cdot b \cdot d \cdot a \in a \cdot P^m \cdot b \cdot P^n \cdot a$. It is clear that $c \cdot b \cdot d \in V(a)$ and $c \cdot b \cdot d \in P^m \cdot b \cdot P^n$. Note that if $m > 1$ and $n > 1$, the element $c \cdot b \cdot d$ is in the set $P \cdot (P^{m-1} \cdot b \cdot P^{n-1}) \cdot P \subseteq P \cdot S \cdot P$; if $m = 1$ and $n = 1$, the element $c \cdot b \cdot d$ is in the set $P \cdot b \cdot P \subseteq P \cdot S \cdot P$; if $m > 1$ and $n = 1$, then $c \cdot b \cdot d \in P \cdot P^{m-1} \cdot b \cdot P \subseteq P \cdot s \cdot P$, and if $m = 1$ and $n > 1$, then $c \cdot b \cdot d \in P \cdot b \cdot P^{n-1} \cdot P \subseteq P \cdot S \cdot P$. Hence, $V(a) \cap P \cdot S \cdot P \neq \emptyset$. Therefore, a is regular in the semihypergroup (S, P). □

Corollary 2.3.14. *Let S be a regular semigroup and P be a non-empty subset of S such that for each element $a \in S$ there exists some natural numbers m, n such that $a \in a \cdot P^m \cap P^n \cdot a$, then the semihypergroup (S, P) is regular.*

Remark 11. An element a of a semigroup S is called an *r-potent* if $a^r = a$, where $r \geq 2$ is a (smallest) natural number. In particular, an element $a \in S$ is called an *idempotent element* if $a^2 = a$.

Corollary 2.3.15. *If $a \in P$ is an r-potent, then a is regular in the semihypergroup (S, P).*

Proof. Since $a \in P$ and $a = a^r$, $r \geq 2$, it follows that a is regular and $a \in a \cdot P^{r-1} \cap P^{r-1} \cdot a$. Therefore, a is a regular element of the semihypergroup (S, P). □

Corollary 2.3.16. *Let P be the set of all idempotent elements of S. If $a \in S$ is regular in S, then a is regular in the semihypergroup (S, P).*

Proof. Since a is a regular in S, there exists an element $x \in S$ such that $a = a \cdot x \cdot a$. Clearly, the elements $a \cdot x$ and $x \cdot a$ are idempotent elements of S. Therefore, the elements $a \cdot x$ and $x \cdot a$ are in P. Since $a = a \cdot x \cdot a = a \cdot (x \cdot a) = (a \cdot x) \cdot a$, we have $a \in a \cdot P \cap P \cdot a$. Therefore, a is regular in the semihypergroup (S, P). □

Corollary 2.3.17. *Let S be a regular semigroup and P be the set of all idempotent elements of S. Then, (S, P) is a regular semihypergroup.*

Corollary 2.3.18. *Suppose S has a left identity f and a right identity g, and P contains f and g. If $a \in S$ is regular in S, then a is regular in the semihypergroup (S, P).*

Proof. Suppose that $a \in S$. Since $a = f \cdot a$ and $a = a \cdot g$, it follows that $a \in P \cdot a \cap a \cdot P$. Hence, a is a regular element of the semihypergroup (S, P). □

Corollary 2.3.19. *Suppose that S has a left identity f and a right identity g and P contains f and g. If S is regular, then the semihypergroup (S, P) is regular.*

Corollary 2.3.20. *If S is regular with the identity element e and P contains the identity element, then the semihypergroup (S, P) is regular.*

Theorem 2.3.21. *Let P be a right ideal and Q a left ideal of S such that $P \cap Q \neq \emptyset$. If a is regular in (S, P) and in (S, Q), then a is regular in the semihypergroup $(S, P \cap Q)$.*

Proof. Suppose that $a \in S$. Then, $a \in a \cdot P \cdot x \cdot P \cdot a$ for some $x \in P$, and $a \in a \cdot Q \cdot y \cdot Q \cdot a$ for some $y \in Q$. That is, $a = a \cdot p_1 \cdot x \cdot p_2 \cdot a$, where p_1, p_2 are elements of P and $a = a \cdot q_1 \cdot y \cdot q_2 \cdot a$, where q_1, q_2 are elements of Q.

Now, we have

$$\begin{aligned} a = a \cdot p_1 \cdot x \cdot p_2 \cdot a &= (a \cdot p_1 \cdot x \cdot p_2) \cdot (a \cdot q_1 \cdot y \cdot q_2 \cdot a) \\ &= (a \cdot p_1 \cdot x \cdot p_2) \cdot (a \cdot q_1 \cdot y \cdot q_2) \cdot (a \cdot p_1 \cdot x \cdot p_2 \cdot a) \\ &= (a \cdot p_1 \cdot x \cdot p_2) \cdot (a \cdot q_1 \cdot y \cdot q_2) \cdot (a \cdot p_1 \cdot x \cdot p_2) \cdot (a \cdot q_1 \cdot y \cdot q_2 \cdot a) \\ &= a \cdot (p_1 \cdot x \cdot p_2 \cdot a \cdot q_1) \cdot (y \cdot q_2 \cdot a \cdot p_1 \cdot x) \cdot (p_2 \cdot a \cdot q_1 \cdot y \cdot q_2) \cdot a. \end{aligned}$$

Since P is a right ideal and Q is a left ideal of the semigroup S, it follows that $p_1 \cdot x \cdot p_2 \cdot a \cdot q_1 \in P \cap Q$ and $p_2 \cdot a \cdot q_1 \cdot y \cdot q_2 \in P \cap Q$. Therefore, a is regular in $(S, P \cap Q)$. □

Corollary 2.3.22. *Let P be a right ideal and Q be a left ideal of S such that $P \cap Q \neq \emptyset$. If the semihypergroups (S, P) and (S, Q) are regular, then the semihypergroup $(S, P \cap Q)$ is regular.*

Theorem 2.3.23. *Let S be a regular semigroup with the identity element e. Then, the semihypergroup (S, P) is regular if and only if P has a right invertible element and a left invertible element of the semigroup S.*

Proof. Assume that (S, P) is a regular semihypergroup. Then, e is regular in (S, P) and hence there exists an element $y \in S$ such that $e \in e \circ_P y \circ_P e$. That is, $e = e \cdot p \cdot y \cdot q \cdot e = p \cdot y \cdot q$, where p, q are some elements of P. Hence, $p \in P$ is right invertible and $q \in P$ is left invertible.

Conversely, suppose that the subset P of S has a left invertible element x and a right invertible element y. Then, there exist elements $s, t \in S$ such that $s \cdot x = e$ and $y \cdot t = e$. Let $a \in S$. Since S is regular, it follows that there exists an element $b \in S$ such that $a = a \cdot b \cdot a$. Now, we have

$$a = a \cdot b \cdot a = a \cdot e \cdot b \cdot e \cdot a = a \cdot y \cdot t \cdot b \cdot s \cdot x \cdot a \in a \cdot P \cdot (t \cdot b \cdot s) \cdot P \cdot a = a \circ_P (t \cdot b \cdot s) \circ_P a.$$

Hence, the semihypergroup (S, P) is regular. □

Corollary 2.3.24. *Let S be a semigroup with the identity element e and each one-sided invertible element in S is invertible. Then the semihypergroup (S, P) is regular if and only if P has an invertible element of the semigroup S.*

Corollary 2.3.25. *Let P be a non-empty subset of the multiplicative semigroup $M_n(\mathbb{R})$ of all $n \times n$ matrices over $\mathbb{R}$. Then, the semihypergroup $(M_n(\mathbb{R}), P)$ is regular if and only if P has an invertible matrix in the semigroup $M_n(\mathbb{R})$.*

Proof. The proof follows from Corollary 2.3.24, since in the multiplicative semigroup $M_n(\mathbb{R})$, each one-sided invertible matrix is invertible. □

2.4 SUBSEMIHYPERGROUPS AND HYPERIDEALS

In this section, we study the concept of hyperideals. The concept of hyperideals is an interesting and important subject in semihypergroup theory. Hyperideals of semihypergroups have been studied by Davvaz, Corsini, Shabir, Hasankhani, Hila, Naka, and many others.

Definition 2.4.1. A non-empty subset A of a semihypergroup $(H, \circ)$ is called a *subsemihypergroup* of H if $A \circ A \subseteq A$.

Example 24. Consider the semihypergroup $(H, \circ)$ defined in Example 21(3). If we consider $B = [0, t]$ with $0 \leq t \leq 1$, then B is a subsemihypergroup of H.

Definition 2.4.2. Let $(H, \circ)$ be a semihypergroup. A non-empty subset I of H is called a *right (left) hyperideal* of H if for all $x \in I$ and $h \in H$

$$x \circ h \subseteq I \quad (h \circ x \subseteq I).$$

A non-empty subset I of H is called a *hyperideal* (or *two-sided hyperideal*) if it is both a left hyperideal and right hyperideal.

Green's relations on semihypergroups are studied by Hasankhani [88]. Later on in this section, we study them.

Definition 2.4.3. Let $(H, \circ)$ be a semihypergroup. For every $a \in H$, we define

$$aH = (a \circ H) \cup \{a\},$$

$$Ha = (H \circ a) \cup \{a\},$$

$$HaH = (H \circ a \circ H) \cup Ha \cup aH.$$

The *hyper version of Green's relations* are the equivalence relations $\mathcal{L}$, $\mathcal{R}$, $\mathcal{J}$, and $\mathcal{H}$ defined for all $x, y \in H$ by

$$x\mathcal{L}y \Leftrightarrow Hx = Hy,$$
$$x\mathcal{R}y \Leftrightarrow xH = yH,$$
$$x\mathcal{J}y \Leftrightarrow HxH = HyH,$$
$$\mathcal{H} = \mathcal{L} \cap \mathcal{R}.$$

Also, we consider the relations $\mathcal{L}^{\leq}$, $\mathcal{R}^{\leq}$, and $\mathcal{J}^{\leq}$ defined for all $x, y \in H$ by

$$x\mathcal{L}^{\leq}y \Leftrightarrow Hx \subseteq Hy,$$
$$x\mathcal{R}^{\leq}y \Leftrightarrow xH \subseteq yH,$$
$$x\mathcal{J}^{\leq}y \Leftrightarrow HxH \subseteq HyH.$$

Theorem 2.4.4. *Let $(H, \circ)$ be a semihypergroup and A be a non-empty subset of H. Then, A is a right hyperideal of H if and only if for every $x, y \in H$,*

$$x\mathcal{R}^{\leq}y \quad and \quad y \in A \Rightarrow x \in A. \tag{2.1}$$

Proof. Suppose that Eq. (2.1) holds. For all $x \in A$ and $h \in H$, we show that $x \circ h \subseteq A$. To do this, let z be an arbitrary element of $x \circ h$. Suppose that $u \in zH$ is arbitrary. Then, we have

$$u = z \text{ or } u \in z \circ t \text{ for some } t \in H.$$

If $u = z$, then it follows that $u \in x \circ h$. Thus, $u \in xH$.

If $u \in z \circ t$, then we obtain

$$u \in (x \circ h) \circ t = x \circ (h \circ t) \subseteq x \circ H \subseteq xH.$$

Thus, we conclude that $zH \subseteq xH$. This means that $z\mathcal{R}^{\leq}x$. Since $x \in A$, by Eq. (2.1), it follows that $z \in A$. Therefore, $x \circ h \subseteq A$.

Conversely, suppose that A is a right hyperideal of H, $x\mathcal{R}^{\leq}y$ and $y \in A$. Since A is a right hyperideal, it follows that $yH = y \circ H \cup \{y\} \subseteq A$. Thus, $x \in xH \subseteq yH \subseteq A$. □

Theorem 2.4.5. *Let $(H, \circ)$ be a semihypergroup and A be a non-empty subset of H. Then, A is a left hyperideal of H if and only if for every $x, y \in H$,*

$$x\mathcal{L}^{\leq}y \quad and \quad y \in A \Rightarrow x \in A. \tag{2.2}$$

Proof. Suppose that Eq. (2.2) holds. For all $x \in A$ and $h \in H$, we prove that $h \circ x \subseteq A$. To do this, let z be an arbitrary element of $h \circ x$. Suppose that $u \in Hz$ is arbitrary. Then, we have

$$u = z \text{ or } u \in t \circ z \text{ for some } t \in H.$$

If $u = z$, then we obtain $u \in h \circ x$. So, $u \in Hx$.

If $u \in t \circ z$, then we obtain

$$u \in t \circ (h \circ x) = (t \circ h) \circ x \subseteq H \circ x \subseteq Hx.$$

So, we conclude that $Hz \subseteq Hx$. Thus, $z\mathcal{L}^{\leq}x$. Since $x \in A$, by Eq. (2.2), it follows that $z \in A$. Therefore, $h \circ x \subseteq A$.

Conversely, suppose that A is a left hyperideal of H, $x\mathcal{L}^{\leq}y$ and $y \in A$. Since A is a left hyperideal, it follows that $Hy = H \circ y \cup \{y\} \subseteq A$. Hence, $x \in Hx \subseteq Hy \subseteq A$. □

Theorem 2.4.6. *Let $(H, \circ)$ be a semihypergroup and $a \in H$. Then, aH is the smallest right hyperideal containing a.*

Proof. We have

$$aH \circ H = (a \circ H \cup \{a\}) \circ H = (a \circ H \circ H) \cup (a \circ H) \subseteq a \circ H \subseteq aH.$$

So, aH is a right hyperideal of H. Now, suppose that A is a right hyperideal of H containing a. Since $a \in A$, it follows that $aH = a \circ H \cup \{a\} \subseteq A$. □

Right hyperideals of the form aH are called *principal right hyperideals*.

Theorem 2.4.7. *Let $(H, \circ)$ be a semihypergroup and $a \in H$. Then, Ha is the smallest left hyperideal containing a.*

Proof. We have

$$H \circ Ha = H \circ (H \circ a \cup \{a\}) = H \circ H \circ a \cup H \circ a \subseteq Ha.$$

This implies that Ha is a left hyperideal of H. Now, let A be a left hyperideal of H containing a. Clearly, we have $Ha = H \circ a \cup \{a\} \subseteq A$. □

Left hyperideals of the form Ha are called *principal left hyperideals*.

Theorem 2.4.8. *Let $(H, \circ)$ be a semihypergroup. If $a, b \in H$, then the following conditions are equivalent.*

(1) *$a\mathcal{R}^{\leq}b$;*
(2) *$a \in bH$;*
(3) *For all principal right hyperideals J of H, if $b \in J$, then $a \in J$;*
(4) *For all right hyperideals J of H, if $b \in J$, then $a \in J$.*

Proof. It is straightforward. □

Corollary 2.4.9. *Let $(H, \circ)$ be a semihypergroup. If $a, b \in H$, then the following conditions are equivalent.*

(1) *$a\mathcal{R}b$;*
(2) *$a \in bH$ and $b \in aH$;*
(3) *For all principal right hyperideals J of H, $b \in J$ if and only if $a \in J$;*
(4) *For all right hyperideals J of H, $b \in J$ if and only if $a \in J$.*

Proof. The proof follows from Definition 2.4.3 and Theorem 2.4.8. □

Theorem 2.4.10. *Let* $(H, \circ)$ *be a semihypergroup. If* $a, b \in H$*, then the following conditions are equivalent.*

(1) $a\mathcal{L}^{\leq}b$;

(2) $a \in Hb$;

(3) *For all principal left hyperideals* J *of* H*, if* $b \in J$*, then* $a \in J$;

(4) *For all left hyperideals* J *of* H*, if* $b \in J$*, then* $a \in J$.

Proof. It is straightforward. □

Corollary 2.4.11. *Let* $(H, \circ)$ *be a semihypergroup. If* $a, b \in H$*, then the following conditions are equivalent.*

(1) $a\mathcal{L}b$;

(2) $a \in Hb$ *and* $b \in Ha$;

(3) *For all principal left hyperideals* J *of* H*,* $b \in J$ *if and only if* $a \in J$;

(4) *For all left hyperideals* J *of* H*,* $b \in J$ *if and only if* $a \in J$.

Proof. The proof follows from Definition 2.4.3 and Theorem 2.4.10.

□

Definition 2.4.12. Let $(H, \circ)$ be a semihypergroup and B be a non-empty subset of H. Then, B is called a *bi-hyperideal* of H if the following conditions hold.

(1) B is a subsemihypergroup of H;

(2) $B \circ H \circ B \subseteq B$.

Remark 12. Every left (right) hyperideal is a bi-hyperideal.

Example 25

(1) Consider the semihypergroup $(H, \circ)$ defined in Example 21(3). We know that $B = [0, t]$ with $0 \leq t \leq 1$ is a subsemihypergroup of H. In addition, we have

$$B \circ H \circ B = \left[0, \frac{t^2}{4}\right] \subseteq [0, t] = B.$$

Therefore, B is a bi-hyperideal of H.

(2) Let $H = \{a, b, c, d\}$ be a semihypergroup with the following hyperoperation

$$x \circ y = \begin{cases} \{b, c\} & \text{if } (x, y) = (a, a) \\ \{b, d\} & \text{if } (x, y) \neq (a, a). \end{cases}$$

It is easy to see that $\{b, d\}$ is a bi-hyperideal of H.

Hila and Naka introduced pure hyperradical of a hyperideal in a semihypergroup with zero element. For this purpose, they defined pure, semipure, and other related types of hyperideals and established some of their basic properties in semihypergroups. For the following definitions and results, the main reference is [89].

In what follows, $(H, \circ)$ will denote a semihypergroup with scalar identity 1, which contains a zero element.

Definition 2.4.13. Let $(H, \circ)$ be a semihypergroup. A right hyperideal A of H is called a *right pure right hyperideal* if for each $x \in A$, there is an element $y \in A$ such that $x \in x \circ y$. If A is a two-sided hyperideal with the property that for each $x \in A$, there is an element $y \in A$ such that $x \in x \circ y$, then A is called a *right pure hyperideal. Left pure left hyperideals* and *left pure hyperideals* are defined analogously.

Definition 2.4.14. Let $(H, \circ)$ be a semihypergroup. A right hyperideal A of H is called a *right semipure right hyperideal* if for each $x \in A$, there is an element y belonging to some proper right hyperideal of H such that $x \in x \circ y$. If A is a two-sided hyperideal with the property that for each $x \in A$, there is an element y belonging to a proper hyperideal of H such that $x \in x \circ y$, then A is called a *right semipure hyperideal. Left semipure left hyperideals* and *left semipure hyperideals* can be similarly defined.

Example 26. Let $(H, \circ)$ be a semihypergroup on $H = \{0, 1, x, y, z, t\}$ with the hyperoperation $\circ$ given by the following table.

$\circ$	0	x	y	z	t	1
0	0	0	0	0	0	0
x	0	$\{1, x\}$	$\{x, y, 1\}$	$\{0, 1, x, z\}$	H	x
y	0	$\{0, y\}$	y	$\{y, t, 1\}$	$\{0, 1, y, t\}$	y
z	0	z	$\{z, t\}$	z	$\{z, t\}$	z
t	0	$\{0, t\}$	$\{0, t\}$	$\{0, t\}$	$\{0, t\}$	t
1	0	x	y	z	t	1

It is easy to see that $I_1 = \{0, t\}$ and $I_2 = \{0, z, t\}$ are right pure right hyperideals of H.

Example 27. Let $(H, \circ)$, be a semihypergroup on $H = \{0, 1, a, b, c, d, e, f\}$ with the hyperoperation $\circ$ given by the following table.

$\circ$	0	a	b	c	d	e	f	1
0	0	0	0	0	0	0	0	0
a	0	a	$\{a, b\}$	c	$\{c, d\}$	e	$\{e, f\}$	a
b	0	b	b	d	d	f	f	b
c	0	c	$\{c, d\}$	c	$\{c, d\}$	c	$\{c, d\}$	c
d	0	d	d	d	d	d	d	d
e	0	e	$\{e, f\}$	c	$\{c, d\}$	e	$\{e, f\}$	e
f	0	f	f	d	d	f	f	f
1	0	a	b	c	d	e	f	1

Clearly, $I_1 = \{0, d\}$, $I_2 = \{0, d, f\}$, and $I_3 = \{0, b, d, f\}$ are right pure right hyperideals of H. Also, $I_4 = \{0, c, d\}$ is a two-sided hyperideal of H which is a right (left) pure hyperideal. Moreover, I_4 is a right and left semipure hyperideal. Finally, $I_5 = \{0, c, d, e, f\}$ is a two-sided hyperideal of H which is a right and left pure hyperideal.

Example 28. Let $H = [0, 1]$. Then, H with the hyperoperation $x \circ y = [0, xy]$ is a semihypergroup. Let $t \in [0, 1]$ and $T = [0, t]$. Then, T is a subsemihypergroup. Moreover, T is a two-sided hyperideal of H which is neither right pure nor left pure, but it is left and right semipure.

Proposition 2.4.15. *Let A be a two-sided hyperideal of H. Then, A is right pure if and only for any right hyperideal B, $B \cap A = B \circ A$.*

Proof. Suppose that A is a right pure hyperideal of H. Since B is a right hyperideal of H, it follows that $B \circ A \subseteq B$. Also, since A is a left hyperideal, it follows that $B \circ A \subseteq A$. Hence, $B \circ A \subseteq B \cap A$. Let $x \in B \cap A$. Since A is a right pure hyperideal, there exists $y \in A$ such that $x \in x \circ y$. As $x \in B$ and $y \in A$, $x \circ y \subseteq B \circ A$. Hence, $x \in B \circ A$. This implies that $B \cap A = B \circ A$.

Conversely, assume that $B \cap A = B \circ A$, for any right hyperideal B of H. We show that A is right pure. Let $x \in A$. Then, $x \circ H = x \circ H \cap A = x \circ H \circ A \subseteq x \circ A$. Since $x \in x \circ H$, it follows that $x \in x \circ A$. Hence, there exists $y \in A$ such that $x \in x \circ y$. This proves that A is a right pure hyperideal. □

Corollary 2.4.16. *If A is a right pure hyperideal, then $A = A \circ A$.*

Proposition 2.4.17. (0) *and H are right pure hyperideals of H. Any union and finite intersection of right pure (respectively, semipure) hyperideals is right pure (respectively, semipure).*

Proof. (0) and H are obviously right pure hyperideals. Let I_1 and I_2 be right pure hyperideals and let $x \in I_1 \cap I_2$. Since $x \in I_1$ and I_1 is right pure, it follows that there exists $y_1 \in I_1$ such that $x \in x \circ y_1$. Similarly, there exists $y_2 \in I_2$ such that $x \in x \circ y_2$. Thus, we have $x \in x \circ y_2 \subseteq (x \circ y_1) \circ y_2 = x \circ (y_1 \circ y_2)$. Since $y_1 \circ y_2 \subseteq I_1 \cap I_2$, it follows that $I_1 \cap I_2$ is right pure. The remaining cases of this proposition can be similarly proved. □

It follows from the above proposition that if I is any hyperideal of H, then I contains a largest pure hyperideal, which is in fact the union of all pure hyperideals contained in I (such hyperideals exist, ie, (0)) and hence a pure hyperideal. The largest pure hyperideal contained in I is denoted by $L(I)$. Similarly, each hyperideal I contains a largest semipure hyperideal, denoted by $H(I)$. Also, $L(I)$ (respectively, $H(I)$) is called the *pure* (respectively, *semipure*) *part* of I.

Definition 2.4.18. Let I be a right pure (respectively, semipure) hyperideal of H. Then, I is called *purely* (respectively, *semipurely*) *maximal* if I

is a maximal element in the set of proper right pure (respectively, semipure) hyperideals.

Example 29. In Example 27, $I_4 = \{0, c, d\}$, $I_5 = \{0, c, d, e, f\}$, and $I_6 = \{0, a, b, c, d, e, f\}$ are right pure hyperideals of the semihypergroup H, and it is clear that I_6 is a purely maximal hyperideal of H.

Definition 2.4.19. Let I be a right pure (respectively, semipure) hyperideal of H. Then, I is called *purely* (respectively, *semipurely*) *prime* if it is proper and if for any right pure (respectively, semipure) hyperideals I_1 and I_2, $I_1 \cap I_2 \subseteq I$ implies $I_1 \subseteq I$ or $I_2 \subseteq I$.

Example 30. In Example 27, the hyperideal I_6 mentioned above is a purely prime hyperideal.

The following propositions are stated for pure and semipure hyperideals simultaneously. However, the proofs are given only for one case, since the proofs are similar for the remaining cases.

Proposition 2.4.20. *Any purely (respectively, semipurely) maximal hyperideal is purely (respectively, semipurely) prime.*

Proof. Suppose that I is purely maximal, and I_1, I_2 are right pure hyperideals such that $I_1 \cap I_2 \subseteq I$. Suppose that $I_1 \nsubseteq I$. Then, $I_1 \cup I = H$. Now, we have

$$I_2 = I_2 \cap H = I_2 \cap (I_1 \cup I) = (I_2 \cap I_1) \cup (I_2 \cap I) \subseteq I \cup I = I.$$

□

Proposition 2.4.21. *The pure (respectively, semipure) part of any maximal hyperideal is purely (respectively, semipurely) prime.*

Proof. Let M be a maximal hyperideal of H. We show that $L(M)$, the pure part of M, is purely prime. Suppose that $I_1 \cap I_2 \subseteq L(M)$ with I_1, I_2 pure. If $I_1 \subseteq M$, then $I_1 \subseteq L(M)$ and we are done. Suppose that $I_1 \nsubseteq M$. Then, $I_1 \cup M = H$. Hence, we have

$$I_2 = I_2 \cap H = I_2 \cap (I_1 \cup M) = (I_2 \cap I_1) \cup (I_2 \cap M) \subseteq M \cup M = M.$$

Hence, $I_2 \subseteq M$. This implies that $I_2 \subseteq L(M)$, since I_2 is pure. □

Proposition 2.4.22. *If I is a right pure (respectively, semipure) hyperideal of H and $a \notin I$, then there exists a purely (respectively, semipurely) prime hyperideal J such that $I \subseteq J$ and $a \notin J$.*

Proof. Consider the set X ordered by inclusion,

$$X = \{J \mid J \text{ is a right semipure hyperideal}, I \subseteq J, a \notin J\}.$$

Then, $X \neq \emptyset$, since $I \in X$. Let $(J_k)_{k \in K}$ be a totally ordered subset of X. Clearly, $\bigcup_k J_k$ is a semipure hyperideal with $a \notin \bigcup_k J_k$. Hence, X is inductively ordered. Therefore, by Zorn's lemma, X has a maximal element J.

We shall show that J is semipurely prime. Suppose that I_1, I_2 are right semipure hyperideals such that $I_1 \nsubseteq J$ and $I_2 \nsubseteq J$. Since I_k $(k = 1, 2)$ and J are semipure, it follows that $I_k \cup J$ is a semipure hyperideal such that $J \subseteq I_k \cup J$. Then, we claim that an $a \in I_k \cup J$ $(k = 1, 2)$. Because, if $a \notin I_k \cup J$, then by the maximality of J, we have $I_k \cup J \subseteq J$. But this contradicts the assumption $I_k \nsubseteq J$ $(k = 1, 2)$. Hence, $a \in (I_1 \cap I_2) \cup J$. Since $a \notin J$, it follows that $I_1 \cap I_2 \nsubseteq J$. Hence, by contrapositivity, we conclude that J is semipurely prime. □

Proposition 2.4.23. *Let I be a proper right pure (respectively, semipure) hyperideal of H. Then, I is contained in a purely (respectively, semipurely) maximal hyperideal.*

Proof. Consider the set, ordered by inclusion,

$$X = \{J | J \text{ is a proper right semipure hyperideal and } J \subseteq I\}.$$

Clearly, $X \neq \emptyset$, since $I \in X$. Moreover, any $J \in X$ is a proper hyperideal. Hence, any directed union of elements in X is still in X. So, X is inductively ordered. Hence, by Zorn's lemma, X contains a maximal element J and any proper semipure hyperideal containing J also contains I and so it belongs to X. But such a hyperideal will be J itself, since J is maximal in X. Therefore, J is semipurely maximal. □

Proposition 2.4.24. *Let I be a proper right pure (respectively, semipure) hyperideal of H. Then, I is the intersection of purely (respectively, semipurely) prime hyperideals of H containing I.*

Proof. By Propositions 2.4.23 and 2.4.20, there exists a set

$$\{P_\alpha | P_\alpha \text{ is a semipurely prime hyperideal containing } I,\ \alpha \in \Lambda\}.$$

Hence, $I \subseteq \bigcap_{\alpha \in \Lambda} P_\alpha$. In order to prove $\bigcap_{\alpha \in \Lambda} P_\alpha \subseteq I$, assume that there exists an element x such that $x \notin I$. Then, by Proposition 2.4.22, there exists a semipurely prime hyperideal P_{α_0} such that $I \subseteq P_{\alpha_0}$, but $x \notin P_{\alpha_0}$. This implies that $x \notin \bigcap_{\alpha \in \Lambda} P_\alpha$. This proves the proposition. □

Definition 2.4.25. Let A be a right hyperideal of H and let $\{K_\alpha | \alpha \in \Lambda\}$ be the set of right pure hyperideals containing A. Then, we define $P(A) = \bigcap_{\alpha \in \Lambda} K_\alpha$ and call it the *pure hyperradical* of A. Note that the set $\{K_\alpha | K_\alpha \text{ is a right pure hyperideal containing } A\}$ is non-empty, since H itself belongs to this set.

Proposition 2.4.26. *If $P(A)$ is the pure hyperradical of the hyperideal A, then each of the following statement holds.*

(1) *$P(A)$ is either pure or semipure hyperideal containing A.*

(2) *$P(A)$ is contained in every right pure hyperideal that contains A.*

(3) *If P_α are those purely prime hyperideals that contain A, then $P(A) = \bigcap_\alpha P_\alpha$.*

Proof. (1) If the set

$$\{K_\alpha | \alpha \in \Lambda,\ K_\alpha \text{ is a right pure hyperideal of } H \text{ containing } A\}$$

consists of H alone, then $P(A) = H$. Hence, $P(A)$ is pure in this case. If the set $\{K_\alpha | K_\alpha$ is a right pure hyperideal containing $A\}$ has only a finite number of elements, then $P(A)$ is pure by Proposition 2.4.17. In general, $P(A)$ is semipure.

(2) This is obvious.

(3) Since every pure hyperideal is contained in a purely maximal hyperideal (Proposition 2.4.23) and every purely maximal hyperideal is purely prime (Proposition 2.4.20), the set $\{P_\alpha | P_\alpha$ is purely prime containing $A\}$ is non-empty. Hence, from Part (2), it follows that $P(A) \subseteq \bigcap_\alpha P_\alpha$. We prove that $\bigcap_\alpha P_\alpha \subseteq P(A)$. To prove this, assume that $x \notin P(A)$. We have $P(A) = \bigcap_\alpha K_\alpha$, where each K_α is a right pure hyperideal containing A. Hence, $x \notin K_{\alpha_0}$ for some α_0. Thus, K_{α_0} is a proper pure hyperideal which contains A but misses x. Hence, by Proposition 2.4.22, there exists a purely prime hyperideal P_{α_0} such that $A \subseteq K_{\alpha_0} \subseteq P_{\alpha_0}$ and $x \notin P_{\alpha_0}$. Hence, $x \notin \bigcap_\alpha P_\alpha$, where P_αs are purely prime hyperideals containing A. From this we conclude that $P(A) = \bigcap_\alpha P_\alpha$. □

Definition 2.4.27. Let A be a hyperideal of H. Then, $H(A)$ is the semipure part of A, that is, the union of all semipure hyperideals contained in A is called the *semipure hyperradical* of A.

Proposition 2.4.28. *For each hyperideal A, $H(A)$ is the intersection of semipurely prime hyperideals.*

Proof. It follows from Proposition 2.4.24. □

Note that $P(A)$ and $H(A)$ are distinct in general. For example, if $A = \{0,\ d\}$ of Example 27, then $P(A) = \{0,\ c,\ d\}$ and $H(A) = \{0\}$.

2.5 QUASI-HYPERIDEALS

In [61], Hila, Davvaz, and Naka introduced the notion of quasi-hyperideal in semihypergroups. Moreover, they studied the notion of an (m, n)-quasi-hyperideal, n-right hyperideal, and m-left hyperideal in semihypergroups and relations between them. In this section, we study these concepts. The main reference for this section is [61].

Definition 2.5.1. Let $(H, \circ)$ be a semihypergroup and Q be a non-empty subset of H. Then, Q is called a *quasi-hyperideal* of H if

$$Q \circ H \cap H \circ Q \subseteq Q.$$

Example 31. Let $(H, \circ)$ be a semihypergroup on $H = \{x,\ y,\ z,\ t\}$ with the hyperoperation $\circ$ given by the following table.

$\circ$	x	y	z	t
x	x	$\{x, y\}$	$\{x, z\}$	H
y	y	y	$\{y, t\}$	$\{y, t\}$
z	z	$\{z, t\}$	z	$\{z, t\}$
t	t	t	t	t

It is easy to see that $Q_1 = \{t\}$, $Q_2 = \{z,\ t\}$, $Q_3 = \{y,\ t\}$, and $Q_4 = \{y,\ z,\ t\}$ are quasi-hyperideals of H.

Example 32. Let $(H, \circ)$ be a semihypergroup on $H = \{a,\ b,\ c,\ d,\ e,\ f\}$ with the hyperoperation $\circ$ given by the following table.

$\circ$	a	b	c	d	e	f
a	a	$\{a, b\}$	c	$\{c, d\}$	e	$\{e,f\}$
b	b	b	d	d	f	f
c	c	$\{c, d\}$	c	$\{c, d\}$	c	$\{c, d\}$
d	d	d	d	d	d	d
e	e	$\{e,f\}$	c	$\{c, d\}$	e	$\{e,f\}$
f	f	f	d	d	f	f

Clearly, $Q_1 = \{d\}$, $Q_2 = \{d,\ f\}$, and $Q_3 = \{b,\ d,\ f\}$ are quasi-hyperideals of H.

Example 33. Let $H = [0, 1]$. Then, H with the hyperoperation $x \circ y = [0, xy]$ is a semihypergroup. Let $t \in [0, 1]$ and $T = [0, t]$. Then, T is a semihypergroup, and moreover, T is a quasi-hyperideal of H.

Example 34. Let $H = (0, 1)$. Then, H with the hyperoperation $x \circ y = \{\frac{xy}{2^k} | 0 \leq k \leq 1\}$ is a semihypergroup. Let $Q_i = (0, \frac{1}{2^i})$, where $i \in \mathbb{N}$. Then, Q_i is a semihypergroup, and moreover, Q_i is a quasi-hyperideal of H.

Example 35. Let $H = \mathbb{N}$. Then, H with the hyperoperation $x \circ y = x + y + 5\mathbb{N}$ is a semihypergroup. Let $Q_i = \{i,\ i+1, \ldots\}$, where $i \in \mathbb{N}$. Then, Q_i are semihypergroups, and moreover, Q_i are quasi-hyperideals of H.

Definition 2.5.2. Let $(H, \circ)$ be a semihypergroup and Q be a subsemihypergroup of H. Then, Q is called an (m, n)*-quasi-hyperideal* of H if $H^m \circ Q \cap Q \circ H^n \subseteq Q$, where m and n are positive integers.

Example 36. All the quasi-hyperideals of the Examples 31, 32, 33, 34, and 35 are (m, n)-quasi-hyperideals of H, respectively.

It is clear that a quasi-hyperideal Q of a semihypergroup H is a $(1, 1)$-quasi-hyperideal of H. Moreover, an (m, n)-quasi-hyperideal of H is a

(k, l)-quasi-hyperideal of H for all $k \geq m$ and $l \geq n$. The following example shows that an (m, n)-quasi-hyperideal of a semihypergroup H need not be a quasi-hyperideal of H.

Example 37. Let $H = \mathbb{N}$. Then, H with the hyperoperation $x \circ y = x + y + 5\mathbb{N}$ is a semihypergroup. Let $Q = \{5\} \cup \{k \in \mathbb{N} | k \geq 12\}$. Then, Q is a semihypergroup. It can be easily seen that Q is a $(2, 3)$-quasi-hyperideal of H. Since $5 \circ 1 \subseteq \mathbb{N} \circ Q \cap Q \circ \mathbb{N} \nsubseteq Q$, it follows that Q is not a quasi-hyperideal of H.

If we consider the class of quasi-hyperideals in semihypergroup, we observe that it is the generalization of the class of one-sided hyperideals in semihypergroups. It is clear that every one-sided hyperideal of a semihypergroup is a quasi-hyperideal of H.

Lemma 2.5.3. *Let $(H, \circ)$ be a semihypergroup and A_i a subsemihypergroup of H for all $i \in I$. If $\bigcap_{i \in I} A_i \neq \emptyset$, then $\bigcap_{i \in I} A_i$ is a subsemihypergroup of H.*

Proof. Assume that $\bigcap_{i \in I} A_i \neq \emptyset$. Let $a, b \in \bigcap_{i \in I} A_i$. Then, $a, b \in A_i$ for all $i \in I$. Since A_i is a subsemihypergroup of H for all $i \in I$, $a \circ b \subseteq A_i$ for all $i \in I$. Hence, $a \circ b \subseteq \bigcap_{i \in I} A_i$. Thus, $\bigcap_{i \in I} A_i$ is a subsemihypergroup of H. □

Proposition 2.5.4. *Let $(H, \circ)$ be a semihypergroup. Let Q and A be an (m, n)-quasi-hyperideal and subsemihypergroup of H, respectively, for all $i \in I$. Then, $A \cap Q$ is either empty or an (m, n)-quasi-hyperideal of A.*

Proof. If $A \cap Q$ is not empty, then $A \cap Q$ is a subset of A, such that $((A \cap Q) \circ A^m) \cap (A \cap Q) \subseteq A \circ A \subseteq A$ and $((A \cap Q) \circ A^m) \cap (A \circ (A \cap Q)) \subseteq Q \circ H^m \cap H^n \circ Q \subseteq Q$. This shows that $A \cap Q$ is an (m, n)-quasi-hyperideal of A. □

Proposition 2.5.5. *Let $(H, \circ)$ be a semihypergroup and $\{Q_i, i \in I\}$ be a set of (m, n)-quasi-hyperideals of H. If $\bigcap_{i \in I} Q_i \neq \emptyset$, then $\bigcap_{i \in I} Q_i$ is an (m, n)-quasi-hyperideal of H.*

Proof. Let Q_i be an (m, n)-quasi-hyperideal of H for $i \in I$. Assume that $\bigcap_{i \in I} Q_i \neq \emptyset$. Then, for every Q_j for all $j \in I$, we have $(H^m \circ (\bigcap_{i \in I} Q_i)) \cap ((\bigcap_{i \in I} Q_i) \circ H^n) \subseteq H^m \circ Q_j \cap Q_j \circ H^n \subseteq Q_j$. This shows that $\bigcap_{i \in I} Q_i$ is an (m, n)-quasi-hyperideal of H. □

Now, let $(H, \circ)$ be a semihypergroup and A be a non-empty subset of H. We denote $\mathcal{F} = \{Q | Q$ as an (m, n)-quasi-hyperideal of H containing $A\}$. It is clear that $\mathcal{F}$ is not empty, since $H \in \mathcal{F}$. Let $\langle A \rangle_{q(m,n)} = \bigcap_{Q \in \mathcal{F}} Q$. It is clear that $\langle A \rangle_{q(m,n)}$ is non-empty since $A \subseteq \langle A \rangle_{q(m,n)}$. Now, $\langle A \rangle_{q(m,n)}$ is an (m, n)-quasi-hyperideal of H and moreover, it is the smallest (m, n)-quasi-hyperideal of H containing A. The (m, n)-quasi-hyperideal $\langle A \rangle_{q(m,n)}$ is called the *(m, n)-quasi-hyperideal of H generated by A.*

Theorem 2.5.6. *Let $(H, \circ)$ be a semihypergroup and $\emptyset \neq A \subseteq H$. Then,*

$$\langle A \rangle_{q(m,n)} = \left(\bigcup_{i=1}^{\max\{m,n\}} A^i \right) \cup (H^m \circ A \cap A \circ H^n).$$

Proof. Let $k = \max\{m, n\}$ and $Q = (\bigcup_{i=1}^{k} A^i) \cup (H^m \circ A \cap A \circ H^n)$. It is clear that $A \subseteq Q$. Let $x, y \in Q$. We have the following cases:

Case 1. $x, y \in \bigcup_{i=1}^{k} A^i$. Then, $x \circ y \subseteq A^t$ for some positive integer t. If $t \leq k$, then $x \circ y \subseteq \bigcup_{i=1}^{k} A^i$. If $t > k$, then $x \circ y \subseteq H^m \circ A \cap A \circ H^n$.

Case 2. $x \in H^m \circ A \cap A \circ H^n$ or $y \in H^m \circ A \cap A \circ H^n$. It can be easily seen that $x \circ y \subseteq H^m \circ A \cap A \circ H^n$. Therefore, $x \circ y \subseteq Q$. Then, Q is a subsemihypergroup of H containing A. We have

$$\begin{aligned} H^m \circ Q \cap Q \circ H^n &= H^m \circ \left(\left(\bigcup_{i=1}^{k} A^i \right) \cup (H^m \circ A \cap A \circ H^n) \right) \\ &\quad \cap \left(\left(\bigcup_{i=1}^{k} A^i \right) \cup (H^m \circ A \cap A \circ H^n) \right) \circ H^n \\ &\subseteq H^m \circ \left(\left(\bigcup_{i=1}^{k} A^i \right) \cup H^m \circ A \right) \\ &\quad \cap \left(\left(\bigcup_{i=1}^{k} A^i \right) \cup A \circ H^n \right) \circ H^n \\ &\subseteq H^m \circ A \cap A \circ H^n \subseteq Q. \end{aligned}$$

Thus, we conclude that Q is an (m, n)-quasi-hyperideal of H containing A.

Now, we show that Q is the smallest. Let Q' be any (m, n)-quasi-hyperideal of H containing A. Then, $A^i \subseteq Q'$ for all positive integers i and $H^m \circ A \cap A \circ H^n \subseteq H^m \circ Q' \cap Q' \circ H^n \subseteq Q'$. Therefore, $Q = (\bigcup_{i=1}^{k} A^i) \cup (H^m \circ A \cap A \circ H^n) \subseteq Q'$. Hence, Q is the smallest (m, n)-quasi-hyperideal of H containing A. Therefore, we obtain the requested result. □

Example 38. Let H be the semihypergroup defined in Example 32 and $A = \{e, f\} \subseteq H$. Then, applying the result of the above Theorem, we obtain that $\langle A \rangle_{q(m,n)} = \{c, d, e, f\}$.

Let $(H, \circ)$ be a semihypergroup and L a subsemihypergroup of H. Then, L is called an *m-left hyperideal* of H if $H^m \circ L \subseteq L$, where m is any positive integer. Dually, if $R \circ H^n \subseteq R$, then R is called an *n-right hyperideal* of H, where n is any positive integer. In the following theorems we prove some results concerning m-left and n-right hyperideal of semihypergroups.

Theorem 2.5.7. *Let $(H, \circ)$ be a semihypergroup. The following statements hold.*

(1) *Let L_i be an m-left hyperideal of H for all $i \in I$. If $\bigcap_{i \in I} L_i \neq \emptyset$, then $\bigcap_{i \in I} L_i$ is an m-left hyperideal of H.*

(2) *Let R_i be an n-right hyperideal of H for all $i \in I$. If $\bigcap_{i \in I} R_i \neq \emptyset$, then $\bigcap_{i \in I} R_i$ is an n-right hyperideal of H.*

Proof. (1) Assume that $\bigcap_{i \in I} L_i \neq \emptyset$. Let $a \in H^m \circ (\bigcap_{i \in I} L_i)$. It follows that $a \in x \circ l$ for some $x \in H^m$ and $l \in \bigcap_{i \in I} L_i$. Then, $l \in L_i$ for all $i \in I$. Then, $a \in H^m \circ L_i$ for all $i \in I$. Since L_i is an m-left hyperideal of H for all $i \in I$, $a \in L_i$ for all $i \in I$. Thus, $a \in \bigcap_{i \in I} L_i$. Therefore, $\bigcap_{i \in I} L_i$ is an m-left hyperideal of H.

(2). It can be proved in the similar way with (1). □

Now, let $(H, \circ)$ be a semihypergroup and A be a non-empty subset of H. We denote $\mathcal{F} = \{L | L$ as an m-left hyperideal of H containing $A\}$. It is clear that $\mathcal{F}$ is not empty since $H \in \mathcal{F}$. Let $(A)_{l(m)} = \bigcap_{L \in \mathcal{F}} L$. It is clear that $(A)_{l(m)}$ is non-empty since $A \subseteq (A)_{l(m)}$. By Theorem 2.5.7(1), $(A)_{l(m)}$ is an m-left hyperideal of H and moreover, it is the smallest m-left hyperideal of H containing A. The m-left hyperideal $(A)_{l(m)}$ is called the *m-left hyperideal of H generated by A*. The *n-right hyperideal $(A)_{r(n)}$ of H generated by A* is defined analogously.

Theorem 2.5.8. *Let $(H, \circ)$ be a semihypergroup and $\emptyset \neq A \subseteq H$. The following statements hold.*

(1) $(A)_{l(m)} = \left(\bigcup_{i=1}^{m} A^i\right) \cup H^m \circ A$.

(2) $(A)_{r(n)} = \left(\bigcup_{i=1}^{n} A^i\right) \cup A \circ H^n$.

Proof. It is similar to the proof of Theorem 2.5.6. □

Theorem 2.5.9. *Let $(H, \circ)$ be a semihypergroup and L, R be an m-left hyperideal and n-right hyperideal of H, respectively. Then, $L \cap R$ is an (m, n)-quasi-hyperideal of H.*

Proof. By the properties of L and R, we have $R^m \circ L^n \subseteq H^m \circ L \cap R \circ H^n \subseteq L \cap R$. Then, $L \cap R$ is non-empty. By Lemma 2.5.3 it follows that $L \cap R$ is a subsemihypergroup of H. Now we have

$$(H^m \circ (L \cap R)) \cap ((L \cap R) \circ H^n) \subseteq H^m \circ L \cap R \circ H^n \subseteq L \cap R,$$

which proves that $L \cap R$ is an (m, n)-quasi-hyperideal of H. □

We say that an (m, n)-quasi-hyperideal Q has the *(m, n) intersection property* if Q is the intersection of an m-left hyperideal and an n-right hyperideal of semihypergroup H. In this case every m-left hyperideal and every n-right hyperideal have the (m, n) intersection property. If any arbitrary family of (m, n)-quasi-hyperideals of H has the (m, n) intersection property, then H is said to have an intersection property of (m, n)-quasi-hyperideals. The following theorem characterizes (m, n)-quasi-hyperideals having the (m, n) intersection property.

Theorem 2.5.10. *Let $(H, \circ)$ be a semihypergroup and Q be a subsemihypergroup of H. Then, the following statements are equivalent.*

(1) *Q is an (m, n)-quasi-hyperideal of H with the (m, n) intersection property.*

(2) $(Q \cup H^m \circ Q) \cap (Q \cup Q \circ H^n) = Q$.

(3) $H^m \circ Q \cap (Q \cup Q \circ H^n) \subseteq Q$.

(4) $Q \circ H^n \cap (Q \cup H^m \circ Q) \subseteq Q$.

Proof. (1 ⇒ 2): Clearly, we have

$$(Q \cup H^m \circ Q) \cap (Q \cup Q \circ H^n) = Q \cup (H^m \circ Q \cap Q \circ H^n) = Q.$$

2 ⇒ 1): We have

$$H^m \circ Q \cap Q \circ H^n \subseteq (Q \cup H^m \circ Q) \cap (Q \cup Q \circ H^n) = Q.$$

So, Q is an (m, n)-quasi-hyperideal of H. We show that $Q \cup H^m \circ Q$ is an m-left hyperideal of H and $Q \cup Q \circ H^n$ is an n-right hyperideal of H. Let $L = Q \cup H^m \circ Q$ and $R = Q \cup Q \circ H^n$. We show first that L is a subsemihypergroup of H. Let $a, b \in L$. We have the following cases:

Case 1. $a, b \in Q$. Since Q is a subsemihypergroup of H, it follows that $a \circ b \subseteq Q \subseteq L$.

Case 2. $a \in Q$ and $b \in H^m \circ Q$. Then, $a \circ b \subseteq Q \circ H^m \circ Q \subseteq H^m \circ Q \subseteq L$.

Case 3. $a \in H^m \circ Q$ and $b \in Q$. Then, $a \circ b \subseteq H^m \circ Q \circ Q \subseteq H^m \circ Q \subseteq L$.

Case 4. $a \in H^m \circ Q$ and $b \in H^m \circ Q$. Then, $a \circ b \subseteq H^m \circ Q \circ H^m \circ Q \subseteq H^m \circ Q \subseteq L$.

Therefore, L is a subsemihypergroup of H. We have

$$H^m \circ L = H^m \circ (Q \cup H^m \circ Q) = H^m \circ Q \cup H^{2m} \circ Q \subseteq H^m \circ Q \subseteq L.$$

Hence, L is an m-left hyperideal of H. In the similar way, R is an n-right hyperideal of H.

(2 ⇒ 3): Let $(Q \cup H^m \circ Q) \cap (Q \cup Q \circ H^n) = Q$. Since $H^m \circ Q \subseteq Q \cup H^m \circ Q$, we have $H^m \circ Q \cap (Q \cup Q \circ H^n) \subseteq (Q \cup H^m \circ Q) \cap (Q \cup Q \circ H^n) = Q$.

(3 ⇒ 2): Let $H^m \circ Q \cap (Q \cup Q \circ H^n) \subseteq Q$. We have

$$Q \subseteq (Q \cup H^m \circ Q) \cap (Q \cup Q \circ H^n). \quad (2.3)$$

Let $x \in (Q \cup H^m \circ Q) \cap (Q \cup Q \circ H^n)$. Since $H^m \circ Q \cap (Q \cup Q \circ H^n) \subseteq Q$, it follows that $x \in Q$. So, we have

$$(Q \cup H^m \circ Q) \cap (Q \cup Q \circ H^n) \subseteq Q. \quad (2.4)$$

By Eqs. (2.3) and (2.4), we have the requested result.

The proofs for (2 ⇒ 4) and (4 ⇒ 2) are similar to the proofs of (2 ⇒ 3) and (3 ⇒ 2), respectively. □

Proposition 2.5.11. *Let* $(H, \circ)$ *be a semihypergroup and* Q *be an* (m, n)*-quasi-hyperideal of* H*. If* $H^m \circ Q \subseteq Q \circ H^n$ *or* $Q \circ H^n \subseteq H^m \circ Q$*, then* Q *has the* (m, n) *intersection property.*

Proof. Assume that $H^m \circ Q \subseteq Q \circ H^n$. Then, $H^m \circ Q = H^m \circ Q \cap Q \circ H^n \subseteq Q$ which shows that Q is an m-left hyperideal of H. Thus, Q has the (m, n) intersection property. In a similar way, if we assume that $Q \circ H^n \subseteq H^m \circ Q$, then Q is an n-right hyperideal of H. In this case also Q has the (m, n) intersection property. □

The following theorem deals with the intersection property of (m, n)-quasi-hyperideals of regular semihyperideals.

Theorem 2.5.12. *Every regular semihypergroup* $(H, \circ)$ *has the intersection property of* (m, n)*-quasi-hyperideals for any positive integer* $m, n \in \mathbb{N}$.

Proof. Let Q be an (m, n)-quasi-hyperideal of a regular semihypergroup H. Then, it can be easily shown that $Q \subseteq Q \circ H^n$. Thus, $Q \cup Q \circ H^n = Q \circ H^n$. Therefore, $H^m \circ Q \cap (Q \cup Q \circ H^n) = H^m \circ Q \cap Q \circ H^n \subseteq Q$. By Theorem 2.5.10, it follows that Q has the intersection property. □

Let H be a semihypergroup and A be a non-empty subset of H. A is called an (m, n)*-hyperideal* if $H^m \circ A \circ H^n \subseteq A$. Then, $(m, 0)$-hyperideal of H is $H^m \circ A \subseteq A$ and $(0, n)$-hyperideal is $A \circ H^n \subseteq A$.

Proposition 2.5.13. *Let* $(H, \circ)$ *be a regular semihypergroup. Then, a non-empty subset* Q *of* H *is an* (m, n)*-quasi-hyperideal of* H *if and only if it is the intersection of a* $(m, 0)$*-hyperideal and a* $(0, n)$*-hyperideal of* H.

Proof. Let $\emptyset \neq Q \subseteq H$ be an (m, n)-quasi-hyperideal of H, ie, $Q^m \circ H \cap H \circ Q^n \subseteq Q$, which is possible only when Q is the intersection of $(m, 0)$-hyperideal and $(0, n)$-hyperideal of H, which is obvious as $Q^m \circ H \cap H \circ Q^n \subseteq Q$.

Conversely, suppose that $Q = L \cap R$, where $H^m \circ L \subseteq L$ and $R \circ H^n \subseteq R$. Then, we have

$$H^m \circ Q = H^m \circ (L \cap R) \subseteq H^m \circ L \subseteq L,$$
$$Q \circ H^n = (L \cap R) \circ H^n \subseteq R \circ H^n \subseteq R.$$

Hence, Q is an (m, n)-quasi-hyperideal of H. □

Definition 2.5.14. Let $(H, \circ)$ be a semihypergroup and L be an m-left hyperideal of H. Then, L is called a *minimal m-left hyperideal* of H if L does not properly contain any m-left hyperideal of H.

Definition 2.5.15. Let $(H, \circ)$ be a semihypergroup and R be an n-right hyperideal of H. Then, R is called a *minimal n-right hyperideal* of H if R does not properly contain any n-right hyperideal of H.

Definition 2.5.16. Let $(H, \circ)$ be a semihypergroup and Q be an (m, n)-quasi-hyperideal of H. Then, Q is called a *minimal (m, n)-quasi-hyperideal* of H if Q does not properly contain any (m, n)-quasi-hyperideal of H.

Lemma 2.5.17. *Let H be a semihypergroup and $a \in H$. The following statements hold.*

(1) *$H^m \circ a$ is an m-left hyperideal of H.*
(2) *$a \circ H^n$ is an n-right hyperideal of H.*
(3) *$H^m \circ a \cap a \circ H^n$ is an (m, n)-quasi-hyperideal of H.*

Proof

(1) We have $H^m \circ (H^m \circ a) \subseteq H^m \circ a$. Hence, (1) holds.
(2) It is similar to (1).
(3) It follows by (1), (2), and Theorem 2.5.9.

□

Theorem 2.5.18. *Let $(H, \circ)$ be a semihypergroup and Q be an (m, n)-quasi-hyperideal of H. Then, Q is minimal if and only if Q is the intersection of some minimal m-left hyperideal L and some minimal n-right hyperideal R of H.*

Proof. Assume that Q is a minimal (m, n)-quasi-hyperideal of H. Let $a \in Q$. By Lemma 2.5.17, it follows that $H^m \circ a$, $a \circ H^n$, and $H^m \circ a \cap a \circ H^n$ are an m-left hyperideal, an n-right hyperideal, and an (m, n)-quasi-hyperideal of H, respectively. By the minimality of Q, since $H^m \circ a \cap a \circ H^n \subseteq H^m \circ Q \cap Q \circ H^n \subseteq Q$, we have $H^m \circ a \cap a \circ H^n = Q$.

We have to show now the minimality of the m-left hyperideal $H^m \circ a$ and the minimality of the n-right hyperideal $a \circ H^n$. Let L be an m-left hyperideal of H contained in $H^m \circ a$. Then, we have $L \cap a \circ H^n \subseteq H^m \circ a \cap a \circ H^n = Q$. By the minimality of Q, since $L \cap a \circ H^n$ is an (m, n)-quasi-hyperideal of H, it follows that $L \cap a \circ H^n = Q$. Then, $Q \subseteq L$. Therefore, $H^m \circ a \subseteq H^m \circ Q \subseteq H^m \circ L \subseteq L$. This implies $L = H^m \circ a$. Thus, the

m-left hyperideal $H^m \circ a$ is minimal. In a similar way, dually the minimality of the n-right hyperideal $a \circ H^n$ can be proved.

Conversely, assume that $Q = L \cap R$ for some minimal m-left hyperideal L and some minimal n-right hyperideal R of H. Let Q' be an (m, n)-quasi-hyperideal of H contained in Q. Then, we have

$$H^m \circ Q' \subseteq H^m \circ Q \subseteq H^m \circ L \subseteq L \quad \text{and} \quad Q' \circ H^n \subseteq Q \circ H^n \subseteq R \circ H^n \subseteq R.$$

It can be easily proved that $H^m \circ Q'$ and $Q' \circ H^n$ are an m-left hyperideal and an n-right hyperideal of H, respectively. The minimality of L and R implies $H^m \circ Q' = L$ and $Q' \circ H^n = R$, hence $Q = L \cap R = H^m \circ Q' \cap Q' \circ H^n \subseteq Q'$. Then, $Q = Q'$. Therefore, Q is a minimal (m, n)-quasi-hyperideal of H. □

The following propositions give necessary and sufficient conditions for the existence of a minimal (m, n)-quasi-hyperideal of a semihypergroup.

An immediate corollary of Theorem 2.5.18 is the following.

Corollary 2.5.19. *Let $(H, \circ)$ be a semihypergroup. Then, H has at least one minimal (m, n)-quasi-hyperideal if and only if H has at least one minimal m-left hyperideal and at least one minimal n-right hyperideal.*

Theorem 2.5.20. *Let $(H, \circ)$ be a semihypergroup. The following statements hold.*

(1) *An m-left hyperideal L is minimal if and only if $H^m \circ a = L$ for all $a \in L$.*

(2) *An n-right hyperideal R is minimal if and only if $a \circ H^n = R$ for all $a \in R$.*

(3) *An (m, n)-quasi-hyperideal Q is minimal if and only if $H^m \circ a \cap a \circ H^n = Q$ for all $a \in Q$.*

Proof. (1) Assume that L is minimal. Let $a \in L$. Then, $H^m \circ a \subseteq H^m \circ L \subseteq L$. By Lemma 2.5.17(1), it follows that $H^m \circ a$ is an m-left hyperideal of H. Since L is a minimal m-left hyperideal of H, we have $H^m \circ a = L$.

Conversely, assume that $H^m \circ a = L$ for all $a \in L$. Let L' be an m-left hyperideal of H contained in L. Let $x \in L' \subseteq L$. Then, $H^m \circ x = L$. We have:

$$L = H^m \circ x \subseteq H^m \circ L' \subseteq L'.$$

This implies that $L = L'$. Therefore, L is minimal.

Statements (2) and (3) can be proved similar to (1). □

Definition 2.5.21. Let $(H, \circ)$ be a semihypergroup. Then, H is called an *m-left simple semihypergroup* if H is a unique m-left hyperideal of H.

Definition 2.5.22. Let $(H, \circ)$ be a semihypergroup. Then, H is called an *n-right simple semihypergroup* if H is a unique n-right hyperideal of H.

Definition 2.5.23. Let $(H, \circ)$ be a semihypergroup. Then, H is called an *(m, n)-quasi-simple semihypergroup* if H is a unique (m, n)-quasi-hyperideal of H.

Theorem 2.5.24. *Let $(H, \circ)$ be a semihypergroup. The following statements hold.*

(1) *H is an m-left simple semihypergroup if and only if $H^m \circ a = H$ for all $a \in H$.*

(2) *H is an n-right simple semihypergroup if and only if $a \circ H^n = H$ for all $a \in H$.*

(3) *H is an (m, n)-quasi-simple semihypergroup if and only if $H^m \circ a \cap a \circ H^n = H$ for all $a \in H$.*

Proof. (1) Since H is an m-left simple semihypergroup, it follows that H is a minimal m-left hyperideal of H. By Theorem 2.5.20(1), $H^m \circ a = H$ for all $a \in H$.

Conversely, assume that $H^m \circ a = H$ for all $a \in H$. By Theorem 2.5.20(1), H is a minimal m-left hyperideal of H and so, H is an m-left simple semihypergroup.

Statements (2) and (3) can be proved similar to (1). □

Theorem 2.5.25. *Let $(H, \circ)$ be a semihypergroup. The following statements hold.*

(1) *If an m-left hyperideal L of H is an m-left simple semihypergroup, then L is a minimal m-left hyperideal of H.*

(2) *If an n-right hyperideal R of H is an n-right simple semihypergroup, then R is a minimal n-right hyperideal of H.*

(3) *If an (m, n)-quasi-hyperideal Q of H is an (m, n)-quasi-simple semihypergroup, then Q is a minimal (m, n)-quasi-hyperideal of H.*

Proof. (1) Let L be an m-left simple semihypergroup. By Theorem 2.5.24(1), we have $L^m \circ a = L$ for all $a \in L$. For every $a \in L$, we have $L = L^m \circ a \subseteq H^m \circ a \subseteq H^m \circ L \subseteq L$. Then, $H^m \circ a = L$ for all $a \in L$. By Theorem 2.5.20(1), we have that L is minimal.

Statements (2) and (3) can be proved similar to (1). □

2.6 PRIME AND SEMIPRIME HYPERIDEALS

Corsini et al. [44] and Lekkoksung [90, 91] studied the notion of prime and semiprime hyperideals of a semihypergroup. In this section, we define prime and semiprime hyperideals of a semihypergroup and characterize those semihypergroups for which each hyperideal is semiprime. The main reference for this section is [44].

Definition 2.6.1. Let $(H, \circ)$ be a semihypergroup. Then, H is called *semisimple* if for each $h \in H$ there exists $x, y, z \in H$ such that $h \in x \circ h \circ y \circ h \circ z$.

Lemma 2.6.2. *Let $(H, \circ)$ be a semihypergroup with identity. Then, the following conditions are equivalent.*

(1) *H is semisimple;*

(2) *$A \cap B = A \circ B$, for all hyperideals A and B of H;*

(3) *$A = A \circ A$, for all hyperideals A of H;*

(4) *$\langle a \rangle = \langle a \rangle \circ \langle a \rangle$, for all $a \in H$.*

Proof. (1 ⇒ 2): Let $a \in A \cap B$, then $a \in A$ and $a \in B$. Since H is semisimple, there exist $x, y, z \in H$ such that

$$a \in x \circ a \circ y \circ a \circ z = (x \circ a \circ y) \circ (a \circ z) \subseteq A \circ B.$$

Thus, $A \cap B \subseteq A \circ B$.

On the other hand, $A \circ B \subseteq A$ (because A is a hyperideal of H) and $A \circ B \subseteq B$ (because B is a hyperideal of H), so we have $A \circ B \subseteq A \cap B$. Hence, $A \cap B = A \circ B$.

(2 ⇒ 3): Take $B = A$, then by hypothesis $A \cap A = A \circ A$. This implies that $A = A \circ A$.

(3 ⇒ 4): Obvious.

(4 ⇒ 1): As $a \in \langle a \rangle = \langle a \rangle \circ \langle a \rangle$, so

$$a \in (H \circ a \circ H) \circ (H \circ a \circ H) = H \circ a \circ (H \circ H) \circ a \circ H \subseteq H \circ a \circ H \circ a \circ H.$$

This implies that $a \in x \circ a \circ y \circ a \circ z$ for some $x, y, z \in H$. Hence, H is semisimple. □

Definition 2.6.3. Let $(H, \circ)$ is a semihypergroup.

(1) A proper hyperideal I of H is called a *prime hyperideal* of H if for all hyperideals A, B of H, $A \circ B \subseteq I$ implies $A \subseteq I$ or $B \subseteq I$.

(2) A proper hyperideal I of H is called a *semiprime hyperideal* of H if for all hyperideals A of H, $A \circ A \subseteq I$ implies $A \subseteq I$.

Definition 2.6.4. A hyperideal I of a semihypergroup H is called an *irreducible hyperideal* if for all hyperideals I_1, I_2 of H, $I_1 \cap I_2 = I$ implies $I_1 = I$ or $I_2 = I$.

Proposition 2.6.5. *A hyperideal A of a semihypergroup H is prime if and only if A is semiprime and irreducible.*

Proof. Suppose that A is a prime hyperideal of H. Then clearly, A is semiprime. Let B and C be any hyperideals of H such that $B \cap C = A$. Since $B \circ C \subseteq B \cap C = A$ and A is a prime hyperideal of H, so $B \subseteq A$ or

$C \subseteq A$. On the other hand, $A \subseteq B$ and $A \subseteq C$ (since $B \cap C = A$). Hence, $B = A$ or $C = A$.

Conversely, let A be an irreducible semiprime hyperideal of H. Let B and C be any hyperideals of H such that $B \circ C \subseteq A$. Since $(B \cap C) \circ (B \cap C) \subseteq B \circ C \subseteq A$ and A is semiprime, so $B \cap C \subseteq A$. But $A \cup (B \cap C) = (A \cup B) \cap (A \cup C) = A$ and A is irreducible, so we have $A \cup B = A$ or $A \cup C = A$. Hence, $B \subseteq A$ or $C \subseteq A$. Thus, A is prime. □

Proposition 2.6.6. *Let I be a hyperideal of H and $a \in H$ such that $a \notin I$. Then, there exists an irreducible hyperideal A of H such that $I \subseteq A$ and $a \notin A$.*

Proof. Suppose that Ω be the collection of all hyperideals of H which contain I but do not contain "a." Then, Ω is non-empty, because $I \in \Omega$. The collection Ω is partially ordered under inclusion. As every totally ordered subset of Ω is bounded above, so by Zorn's lemma, there exists a maximal element, say, A in Ω. We show that A is an irreducible hyperideal of H. Let C and D be two hyperideals of H such that $C \cap D = A$. If both C and D properly contain A, then we obtain a contradiction by the maximality of A. Hence, $C = A$ or $D = A$, that is, A is irreducible. □

In the next theorem we characterize those semihypergroups in which each hyperideal is semiprime.

Theorem 2.6.7. *Let H be a semihypergroup with identity. Then, the following conditions are equivalent.*

(1) *H is semisimple;*
(2) *$A \cap B = A \circ B$, for all hyperideals A and B of H;*
(3) *$A = A \circ A$, for all hyperideals A of H;*
(4) *Each hyperideal of H is semiprime;*
(5) *Each hyperideal of H is the intersection of prime hyperideals of H which contain it.*

Proof. ($1 \Leftrightarrow 2$) and ($2 \Leftrightarrow 3$): Follow from Lemma 2.6.2.

($3 \Rightarrow 4$): Let A and I be hyperideals of H such that $A \circ A \subseteq I$. By hypothesis $A \circ A = A$, so $A \subseteq I$. Hence, each hyperideal of H is semiprime.

($4 \Rightarrow 5$): Let A be a hyperideal of H. Then, obviously A is contained in the intersection of all irreducible hyperideals of H that contain A. If $a \notin A$, then by Proposition 2.6.6, there exists an irreducible hyperideal of H which contains A but does not contain a. Hence, A is the intersection of all irreducible hyperideals of H that contain it. By hypothesis, each hyperideal of H is semiprime, so each hyperideal of H is the intersection of all irreducible semiprime hyperideals of H that contain it. By Proposition 2.6.5, each irreducible semiprime hyperideal of H is prime. Hence, each hyperideal of H is the intersection of prime hyperideals of H that contain it.

(5 ⇒ 3): Let A be a proper hyperideal of H. Then, $A \circ A$ is a hyperideal of H. By hypothesis,

$$A \circ A = \bigcap_{\alpha} \{A_\alpha | A_\alpha \text{ are prime hyperideal of } H \text{ containing } A \circ A\}.$$

This implies that $A \circ A \subseteq A_\alpha$ for each α. Since each A_α is prime, it follows that $A \subseteq A_\alpha$ for each α and hence $A \subseteq \bigcap_\alpha A_\alpha = A \circ A$. But $A \circ A \subseteq A$ always. Hence, $A \circ A = A$. □

The next proposition shows that in a semisimple semihypergroup the concepts of prime hyperideal and irreducible hyperideal coincide.

Proposition 2.6.8. *Let T be a hyperideal of a semisimple semihypergroup H. Then, the following are equivalent:*

(1) *T is a prime hyperideal of H;*

(2) *T is an irreducible hyperideal of H.*

Proof. (1 ⇒ 2): Suppose that T is a prime hyperideal of H. Let A, B be any hyperideals of H such that $A \cap B = T$. Then, $T \subseteq A$ and $T \subseteq B$. Since $A \circ B \subseteq A \cap B$, it follows that $A \circ B \subseteq T$. Since T is prime, it follows that $A \subseteq T$ or $B \subseteq T$. Thus, $A = T$ or $B = T$.

(2 ⇒ 1):) Suppose that T is an irreducible hyperideal of H. Let A, B be any hyperideals of H such that $A \circ B \subseteq T$. Since H is semisimple, it follows that $A \cap B = A \circ B \subseteq T$. Then, $(A \cap B) \cup T = T$. But $(A \cap B) \cup T = (A \cup T) \cap (B \cup T)$. Hence, $(A \cup T) \cap (B \cup T) = T$. Since T is irreducible, it follows that $A \cup T = T$ or $B \cup T = T$. Thus, we have $A \subseteq T$ or $B \subseteq T$. □

Theorem 2.6.9. *The following conditions are equivalent for a semihypergroup $(H, \circ)$.*

(1) *Each hyperideal of H is prime;*

(2) *H is semisimple and the set of hyperideals of H is a chain.*

Proof. (1 ⇒ 2): Suppose that each hyperideal of H is prime. Then, by Theorem 2.6.7, H is semisimple. Let A, B be hyperideals of H. Then, $A \circ B \subseteq A \cap B$. By hypothesis, each hyperideal of H is prime, so $A \cap B$ is prime. Thus, $A \subseteq A \cap B$ or $B \subseteq A \cap B$, that is $A \subseteq B$ or $B \subseteq A$.

(2 ⇒ 1): Suppose that H is semisimple and the set of hyperideals of H is a chain. Let A, B, C be hyperideals of H such that $A \circ B \subseteq C$. Since H is semisimple, it follows that $A \circ B = A \cap B$. Since the set of hyperideals is a chain, so either $A \subseteq B$ or $B \subseteq A$. Thus, either $A \subseteq C$ or $B \subseteq C$. □

Example 39. Let $(S, \cdot)$ be a semigroup. Define a hyperoperation $\circ$ on S by

$$a \circ b = \{a,\ b,\ a \cdot b\},$$

for all $a, b \in S$. It is easy to see that $(S, \circ)$ is a semihypergroup. The only hyperideal of such semihypergroup is itself. Since $S \circ S = S$, it follows that $(S, \circ)$ is semisimple.

Example 40. Let $(S, \cdot, \leq)$ be an ordered semigroup. Define a hyperoperation $\circ$ on S by

$$a \circ b = \{x \in S | x \leq a \cdot b\},$$

for all $a, b \in S$. Then, it is easy to see that $\circ$ is associative and so $(S, \circ)$ is a semihypergroup.

Now, consider the order semigroup $S = \{a, b, c, d, e\}$ with the following multiplication table and order relation

$\cdot$	a	b	c	d	e
a	a	d	a	d	d
b	a	b	a	d	d
c	a	d	c	d	e
d	a	d	a	d	d
e	a	d	c	d	e

$$\leq = \{(a, a), (a, c), (a, d), (a, e), (b, b), (b, d), (b, e), (c, c), (c, e), (d, d), (d, e), (e, e)\}.$$

Then, the hyperoperation $\circ$ is defined in the following table.

$\circ$	a	b	c	d	e
a	a	$\{a, b, d\}$	a	$\{a, b, d\}$	$\{a, b, d\}$
b	a	b	a	$\{a, b, d\}$	$\{a, b, d\}$
c	a	$\{a, b, d\}$	$\{a, c\}$	$\{a, b, d\}$	$\{a, b, c, d, e\}$
d	a	$\{a, b, d\}$	a	$\{a, b, d\}$	$\{a, b, d\}$
e	a	$\{a, b, d\}$	$\{a, c\}$	$\{a, b, d\}$	$\{a, b, c, d, e\}$

Then, $(S, \circ)$ is a semihypergroup and the only hyperideals of S are $\{a, b, d\}$ and S. Both the hyperideals are idempotent. So, $(S, \circ)$ is a semisimple semihypergroup. Also, both the hyperideals are prime.

Example 41. Consider the ordered semigroup $S = \{a, b, c, d\}$ with the following multiplication table and order relation.

$\cdot$	a	b	c	d
a	a	a	a	a
b	a	a	a	a
c	a	a	b	a
d	a	a	b	b

$$\leq = \{(a, a), (b, b), (c, c), (d, d), (a, b)\}.$$

Then, the hyperoperation $\circ$ on S is defined by the following table.

$\circ$	a	b	c	d
a	a	a	a	a
b	a	a	a	a
c	a	a	$\{a, b\}$	a
d	a	a	$\{a, b\}$	$\{a, b\}$

Then, $(S, \circ)$ is a semihypergroup and the hyperideals of $(S, \circ)$ are $\{a\}$, $\{a, b\}$, $\{a, b, c\}$, $\{a, b, d\}$, and S. No hyperideal is prime or semiprime. But the proper hyperideals $\{a, b, c\}$ and $\{a, b, d\}$ are irreducible.

Example 42. Consider the ordered semigroup $S = \{a, b, c, d\}$ with the following multiplication table and order relation.

$\cdot$	a	b	c	d
a	a	a	a	a
b	a	b	c	a
c	a	a	a	a
d	a	d	a	a

$$\leq = \{(a, a), (b, b), (c, c), (d, d), (a, b), (a, c), (a, d)\}.$$

Then, the hyperoperation $\circ$ on S is defined by the following table.

$\circ$	a	b	c	d
a	a	a	a	a
b	a	$\{a, b\}$	$\{a, c\}$	a
c	a	a	a	a
d	a	$\{a, d\}$	a	a

Then, $(S, \circ)$ is a semihypergroup and the hyperideals of $(S, \circ)$ are $\{a\}$, $\{a, c\}$, $\{a, d\}$, $\{a, c, d\}$, and S. The hyperideal $\{a, c, d\}$ is prime and all other hyperideals are neither prime nor semiprime.

2.7 SEMIHYPERGROUP HOMOMORPHISMS

Homomorphisms of algebraic hyperstructures are studied by Dresher, Ore, Krasner, Kuntzmann, Koskas, Jantosciak, Corsini, Freni, Davvaz, and many others. In this section, we study several kinds of homomorphisms. The main references are [23, 69, 92].

Definition 2.7.1. Let $(H_1, \circ)$ and $(H_2, \star)$ be two semihypergroups. A map $f : H_1 \rightarrow H_2$ is called

(1) a *homomorphism* or *inclusion homomorphism* if for all x, y of H_1, we have $f(x \circ y) \subseteq f(x) \star f(y)$;

(2) a *good (strong) homomorphism* if for all x, y of H_1, we have $f(x \circ y) = f(x) \star f(y)$;

(3) an *isomorphism* if it is a one-to-one and onto good homomorphism. If f is an isomorphism, then H_1 and H_2 are said to be *isomorphic*, which is denoted by $H_1 \cong H_2$.

Example 43. Let $H_1 = \{a, b, c\}$ and $H_2 = \{0, 1, 2\}$ be two semihypergroups with the following hyperoperations:

$\circ$	a	b	c
a	a	H_1	H_1
b	H_1	b	b
c	H_1	b	c

$\star$	0	1	2
0	0	H_2	H_2
1	H_2	1	1
2	H_2	1	$\{1, 2\}$

and let $f : H_1 \to H_2$ is defined by $f(a) = 0, f(b) = 1$, and $f(c) = 2$. Clearly, f is an inclusion homomorphism but it is not a good homomorphism.

Lemma 2.7.2. *Let $(H_1, \circ)$ and $(H_2, \star)$ be two semihypergroups and $f : H_1 \to H_2$ be a good homomorphism. Then, Imf is a subsemihypergroup of H_2.*

Proof. For every $a, b \in H_1$, we have $f(a) \star f(b) = f(a \circ b) \subseteq Imf$. □

We employ for simplicity of notation $x_f = f^{-1}(f(x))$ and for a subset A of H_1, $A_f = f^{-1}(f(A)) = \cup\{x_f | x \in A\}$.

Notice that the defining condition for an inclusion homomorphism is equivalent to

$$x \circ y \subseteq f^{-1}(f(x) \star f(y)).$$

It is also clear for an inclusion homomorphism that

$$(x \circ y)_f \subseteq f^{-1}(f(x) \star f(y)).$$

The defining condition for an inclusion homomorphism is also valid for sets. That is, if A, B are non-empty subsets of H_1, then it follows that

$$f(A \circ B) \subseteq f(A) \star f(B).$$

Applying the above relation for $A = x_f$ and $B = y_f$, we obtain

$$x_f \circ y_f \subseteq f^{-1}(f(x) \star f(y))$$

and

$$\left(x_f \circ y_f\right)_f \subseteq f^{-1}(f(x) \star f(y)).$$

Homomorphisms having various types of properties are defined and studied in the literature. Each of these properties can be viewed as a condition on $f^{-1}(f(x) \star f(y))$. We consider four types of homomorphisms in the following definition.

Definition 2.7.3. Let $(H_1, \circ)$ and $(H_2, \star)$ be two semihypergroups and $f : H_1 \to H_2$ be a mapping. Then, given $x, y \in H_1$, f is called a homomorphism of

type 1, if $f^{-1}(f(x) \star f(y)) = \left(x_f \circ y_f\right)_f$;

type 2, if $f^{-1}(f(x) \star f(y)) = (x \circ y)_f$;

type 3, if $f^{-1}(f(x) \star f(y)) = x_f \circ y_f$;

type 4, if $f^{-1}(f(x) \star f(y)) = (x \circ y)_f = x_f \circ y_f$.

Note that $x \circ y \subseteq (x \circ y)_f$, $x \circ y \subseteq x_f \circ y_f$ and that $(x \circ y)_f \subseteq \left(x_f \circ y_f\right)_f$, $x_f \circ y_f \subseteq \left(x_f \circ y_f\right)_f$. Hence, a homomorphism of any type 1 through 4 is indeed an inclusion homomorphism. Observe that a one-to-one homomorphism of H_1 onto H_2 of any type 1 through 4 is an isomorphism.

Proposition 2.7.4. *Let $(H_1, \circ)$ and $(H_2, \star)$ be two semihypergroups, A, B be non-empty subsets of H_1, and $f : H_1 \to H_2$ be a mapping. Then, if f is a homomorphism of*

(1) *type 1, this implies $f^{-1}(f(A) \star f(B)) = \left(A_f \circ B_f\right)_f$;*

(2) *type 2, this implies $f^{-1}(f(A) \star f(B)) = (A \circ B)_f$;*

(3) *type 3, this implies $f^{-1}(f(A) \star f(B)) = A_f \circ B_f$;*

(4) *type 4, this implies $f^{-1}(f(A) \star f(B)) = (A \circ B)_f = A_f \circ B_f$.*

Proof. Each part is established by a straightforward set theoretical argument. □

Proposition 2.7.5. *Let $(H_1, \circ)$ and $(H_2, \star)$ be two semihypergroups and $f : H_1 \to H_2$ be a mapping. Then, f is a homomorphism of*

(1) *type 4 if and only if f is a homomorphism of type 2 and type 3;*

(2) *type 1 iff f is a homomorphism of type 2 or type 3.*

Proof. (1) It is trivial.

(2) Suppose that $x, y \in H_1$ and f is a homomorphism of type 2. Then,

$$(x \circ y)_f \subseteq (x_f \circ y_f)_f \subseteq f^{-1}(f(x) \star f(y)) = (x \circ y)_f.$$

Similarly, if f is a homomorphism of type 3, then

$$x_f \circ y_f \subseteq (x_f \circ y_f)_f \subseteq f^{-1}(f(x) \star f(y)) = x_f \circ y_f.$$

Hence, in either case, f is a homomorphism of type 1. Thus, (2) holds. □

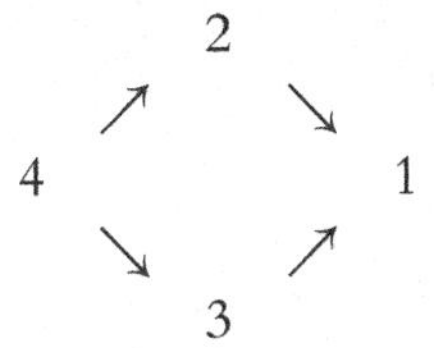

Homomorphism Types

The defining condition for a homomorphism of type 1 or type 2 can easily be simplified if the homomorphism is onto mapping.

Proposition 2.7.6. *Let* $(H_1, \circ)$ *and* $(H_2, \star)$ *be two semihypergroups and* $f : H_1 \to H_2$ *be an onto mapping. Then, given* $x, y \in H_1$, f *is a homomorphism of*

(1) *type 1 if and only if* $f(x_f \circ y_f) = f(x) \star f(y)$;

(2) *type 2 if and only if* $f(x \circ y) = f(x) \star f(y)$.

Proof. It is straightforward. □

Corollary 2.7.7. *Let* $(H_1, \circ)$ *and* $(H_2, \star)$ *be two semihypergroups,* A, B *be non-empty subsets of* H_1*, and* $f : H_1 \to H_2$ *be an onto mapping. Then,* f *is a homomorphism of*

(1) *type 1 implies* $f(A_f \circ B_f) = f(A) \star f(B)$;

(2) *type 2 implies* $f(A \circ B) = f(A) \star f(B)$.

Example 44. Consider $H_1 = \{0, 1, 2\}$ and $H_2 = \{a, b\}$ together with the following hyperoperations:

$\circ$	0	1	2
0	0	$\{0,1\}$	$\{0,2\}$
1	$\{0,1\}$	1	$\{1,2\}$
2	$\{0,2\}$	$\{1,2\}$	2

$\star$	a	b
a	a	$\{a,b\}$
b	$\{a,b\}$	b

and suppose that $f : H_1 \to H_2$ is defined by $f(0) = f(1) = a$ and $f(2) = b$. Then, f is a good homomorphism of type 4.

On a semihypergroup H, we are concerned with equivalence relations for which the family of equivalence classes forms a semihypergroup under the hyperoperation induced by that on H. For an equivalence relation ρ on H, we may use x_ρ, $\overline{x}$, or $\rho(x)$ to denote the equivalence class of $x \in H$. Moreover, generally, if A is a non-empty subset of H, then $A_\rho = \cup\{x_\rho | x \in A\}$.

We let H/ρ (read H modulo ρ) denote the family $\{x_\rho | x \in H\}$ of classes of ρ. The hyperoperation on H induces a hyperoperation $\otimes$ on H/ρ defined by

$$x_\rho \otimes y_\rho = \{z_\rho | z \in x_\rho \circ y_\rho\},$$

where $x, y \in H$. The structure $(H/\rho, \otimes)$ is known as a *factor* or *quotient structure*. Note that in the definition of $\otimes$, the condition $z \in x_\rho \circ y_\rho$ maybe replaced by $z \in (x_\rho \circ y_\rho)_\rho$ or $z_\rho \subseteq (x_\rho \circ y_\rho)_\rho$. Obviously, $\bigcup(x_\rho \otimes y_\rho) = (x_\rho \circ y_\rho)_\rho$.

Proposition 2.7.8. *Let $(H, \circ)$ be a semihypergroup. Then, $(H/\rho, \otimes)$ is a semihypergroup if and only if for all $x, y, z \in H$,*

$$\left((x_\rho \circ y_\rho)_\rho \circ z_\rho\right)_\rho = \left(x_\rho \circ (y_\rho \circ z_\rho)_\rho\right)_\rho.$$

Proof. In H/ρ, we have

$$\begin{aligned}(x_\rho \otimes y_\rho) \otimes z_\rho &= \{u_\rho \mid u \in x_\rho \circ y_\rho\} \otimes z_\rho \\ &= \{t_\rho | t \in u_\rho \circ z_\rho,\ u \in x_\rho \circ y_\rho\} \\ &= \{t_\rho | t \in (x_\rho \circ y_\rho)_\rho \circ z_\rho\}.\end{aligned}$$

Similarly, $x_\rho \otimes (y_\rho \otimes z_\rho) = \{t_\rho | t \in x_\rho \circ (y_\rho \circ z_\rho)_\rho\}$. □

Definition 2.7.9. Let ρ be an equivalence relation on a semihypergroup $(H, \circ)$. Then, given $x, y \in H$, ρ is said to be of

type 1, if H/ρ is a semihypergroup;

type 2, if $x_\rho \circ y_\rho \subseteq (x \circ y)_\rho$;

type 3, if $(x \circ y)_\rho \subseteq x_\rho \circ y_\rho$;

type 4, if $x_\rho \circ y_\rho = (x \circ y)_\rho$.

Observe that being of type 2 is equivalent to $(x_\rho \circ y_\rho)_\rho = (x \circ y)_\rho$, and of type 3 to $(x_\rho \circ y_\rho)_\rho = x_\rho \circ y_\rho$. Note that for an equivalence of type 2, $x_\rho \otimes y_\rho = \{z_\rho | z \in x \circ y\}$.

Proposition 2.7.10. *Let ρ be an equivalence relation on a semihypergroup $(H, \circ)$. Then, ρ is of*

(1) *type 4 if and only if ρ is of type 2 and type 3;*

(2) *type 1 if ρ is of type 2 or type 3.*

Proof. (1) It is straightforward.

(2) Let $x, y, z \in H$. Suppose that ρ is of type 2. Then, we obtain

$$\left((x_\rho \circ y_\rho)_\rho \circ z_\rho\right)_\rho = ((x \circ y) \circ z)_\rho \text{ and } \left(x_\rho \circ (y_\rho \circ z_\rho)_\rho\right)_\rho = (x \circ (y \circ z))_\rho.$$

Suppose that ρ is of type 3. Then, we have

$$\left((x_\rho \circ y_\rho)_\rho \circ z_\rho\right)_\rho = (x_\rho \circ y_\rho) \circ z_\rho \text{ and } \left(x_\rho \circ (y_\rho \circ z_\rho)_\rho\right)_\rho = x_\rho \circ (y_\rho \circ z_\rho).$$

Hence, in either case, Proposition 2.7.8 applies and yields that H/ρ is a semihypergroup. Therefore, ρ is of type 1 and (2) holds. □

Homomorphisms between semihypergroups and equivalence relations on semihypergroups are closely related. The following results are fundamental.

Theorem 2.7.11. *Let $(H_1, \circ)$ and $(H_2, \star)$ be two semihypergroups and $f : H_1 \to H_2$ be an onto mapping. We denote also by f the equivalence relation by f on H_1 whose classes comprise the family $\{x_f | x \in H_1\}$. Then, for $n = 1, 2, 3$, or 4, f is an equivalence relation of type n on H_1 for which H_1/f is canonically isomorphic to H_2 if and only if f is a homomorphism of type n.*

Proof. Let $n = 1$. Suppose that f is a homomorphism of type 1. In order to show f is an equivalence relation of type 1 on H_1, Proposition 2.7.8 is employed. Let $x, y, z \in H_1$. By Corollary 2.7.7 (1),

$$\begin{aligned}\left((x_f \circ y_f)_f \circ z_f\right)_f &= f^{-1}\left(f((x_f \circ y_f)_f \circ z_f)\right)\\ &= f^{-1}(f(x_f \circ y_f) \star f(z))\\ &= f^{-1}(f(x) \star f(y) \star f(z)).\end{aligned}$$

Similarly, we obtain $(x_f \circ (y_f \circ z_f)_f)_f = f^{-1}(f(x) \star f(y) \star f(z))$. Thus, f is an equivalence relation of type 1. The canonical mapping θ of H_1/f onto H_2 is given for $x \in H_1$ by $\theta(x_f) = f(x)$. It is clearly well defined and one-to-one. Moreover, for $x, y \in H_1$,

$$\theta(x_f \otimes y_f) = \theta(\{z_f | z \in x_f \circ y_f\}) = \{f(z) | z \in x_f \circ y_f\} = f(x_f \circ y_f)$$

and

$$\theta(x_f) \star \theta(y_f) = f(x) \star f(y).$$

Therefore, Proposition 2.7.6 (1) yields that H_1/f is canonically isomorphic to H_2 if and only if f is a homomorphism of type 1. The theorem is then established for $n = 1$.

Now, let $n > 1$. If f is a homomorphism of type n, then Proposition 2.7.5 and the theorem for $n = 1$ imply that H_1/f is canonically isomorphic to H_2. On the other hand, if H_1/f is canonically isomorphic to H_2, then the above relations yields for $x, y \in H_1$ that $f^{-1}(f(x) \star f(y)) = (x_f \circ y_f)_f$. Therefore, for $n = 2, 3$, or 4, f is an equivalence relation of type n if and only if f is a homomorphism of type n. □

Corollary 2.7.12. *Let $(H, \circ)$ be a semihypergroup. Let θ be an equivalence relation on H. We denote also by θ the canonical mapping of H onto H/θ. Then, for $n = 1, 2, 3$, or 4, θ is an equivalence relation of type n on H if and only if H/θ is a semihypergroup and θ is a homomorphism of type n of H onto H/θ.*

Proof. By Theorem 2.7.11, if θ is an equivalence relation of type 1, 2, 3, or 4, then H/θ is a semihypergroup. Note that for $x \in H$, $\theta(x) = x_\theta$, an element of H/θ, and that $\theta^{-1}(\theta(x)) = x_\theta$, a subset of H. This compatibility allows the theorem to be applied and yields the corollary. □

Three equivalence relations on a semihypergroup are of such importance that they will be referred to as the fundamental equivalences. They arise as one tries to discriminate between pairs of elements by means of the hyperoperation.

Definition 2.7.13. Let $(H, \circ)$ be a semihypergroup. Let $x, y \in H$. Then, x and y are said to be *operationally equivalent* or *o-equivalent* if $x \circ a = y \circ a$ and $a \circ x = a \circ y$ for every $a \in H$. The elements x and y are said to be *inseparable* or *i-equivalent* if $x \in a \circ b$ when and only when $y \in a \circ b$ for every $a, b \in H$. Also, x and y are said to be *essentially indistinguishable* or *e-equivalent* if they are both operationally equivalent and inseparable.

Obviously, the relations o-, i-, and e-equivalence, denoted respectively by o, i, and e are equivalence relations on a semihypergroup H. For $x \in H$, the o-, i-, and e-equivalence classes of x are hence denoted by x_o, x_i, and x_e, respectively.

Proposition 2.7.14. *Let $(H, \circ)$ be a semihypergroup and $x, y \in H$.*

(1) $x_o \circ y_o = x \circ y$; *o-equivalence is of type* 2.

(2) $(x \circ y)_i = x \circ y$; *i-equivalence is of type* 3.

(3) $x_e = x_o \cap x_i$; $x_e \circ y_e = (x \circ y)_e = x \circ y$; *e-equivalence is of type* 4.

Proof. It is an immediate consequence of definition. □

Corollary 2.7.15. *Given that H is a semihypergroup, H/o, H/i, and H/e are semihypergroups. The canonical mappings of H onto H/o, H/i, and H/e are homomorphisms of type* 2, 3, *and* 4, *respectively.*

Definition 2.7.16. A semihypergroup H in which $x_o = x$ for each $x \in H$, $x_i = x$ for each $x \in H$, or $x_e = x$ for each $x \in H$ is said respectively to be *o-reduced*, *i-reduced*, or *e-reduced*. An e-reduced semihypergroup is simply said to be *reduced*.

Example 45. Let $H = \{a, b, c, d\}$. Let the hyperoperation $\circ$ on H be given by the following table:

$\circ$	a	b	c	d
a	$\{a, b\}$	$\{a, b\}$	$\{c, d\}$	$\{c, d\}$
b	$\{a, b\}$	$\{a, b\}$	$\{c, d\}$	$\{c, d\}$
c	$\{c, d\}$	$\{c, d\}$	a	b
d	$\{c, d\}$	$\{c, d\}$	b	a

One easily checks that $(H, \circ)$ is a semihypergroup. The o-equivalence classes are seen to be $\{a, b\}$, c, and d, whereas the i-equivalence classes are a, b, and $\{c, d\}$. Therefore, the e-equivalence classes are all singletons, that is to say, H is e-reduced.

Proposition 2.7.17. *Let* $(H_1, \circ)$, $(H_2, \star)$, *and* $(H_3, \bullet)$ *be semihypergroups. For* $n = 1, 2, 3$, *or* 4, *let* f *be a homomorphism of type* n *of* H_1 *onto* H_2 *and* g *be a homomorphism of type* n *of* H_2 *onto* H_3. *Then,* gf *is a homomorphism of type* n *of* H_1 *onto* H_3.

Proof. Suppose that $x, y \in H_1$. We first observe that for $z \in H_1$,

$$z_{gf} = f^{-1}g^{-1}gf(z) = f^{-1}(f(z)_g).$$

Let $n = 1$. By the above relation, we obtain

$$gf(x_{gf} \circ y_{gf}) = gf\left(f^{-1}(f(x)_g) \circ f^{-1}(f(y)_g)\right).$$

Since f is onto, Corollary 2.7.7 (1) applies and yields

$$gf\left(f^{-1}(f(x)_g) \circ f^{-1}(f(y)_g)\right) = g(f(x)_g \star f(y)_g).$$

Then, Proposition 2.7.6 (1) gives

$$g\left(f(x)_g \star f(y)_g\right) = gf(x) \bullet gf(y).$$

Hence, gf is a homomorphism of type 1 by the above relations and Proposition 2.7.6 (1).

Let $n = 2$. Similar to the previous case, but simpler.

Let $n = 3$. Since g is of type 3,

$$f^{-1}g^{-1}(gf(x) \bullet gf(y)) = f^{-1}(f(x)_g \star f(y)_g).$$

Since f is onto, Proposition 2.7.4 (3) applies and gives

$$f^{-1}(f(x)_g \star f(y)_g) = f^{-1}(f(x)_g) \circ f^{-1}(f(y)_g).$$

Now, we obtain

$$f^{-1}(f(x)_g) \circ f^{-1}(f(y)_g) = x_{gf} \circ y_{gf}.$$

So, by the above relations, we conclude that gf is a homomorphism of type 3.

Let $n = 4$. Then, gf is a homomorphism of type 4 by Proposition 2.7.5 (1). □

In the following definition, we introduce more types of homomorphisms of semihypergroups that appeared in the literature under various names.

For a, b, we denote $a/b = \{x | a \in x \circ b\}$ and $b \backslash a = \{y | a \in b \circ y\}$.

Definition 2.7.18. Let $(H_1, \circ)$ and $(H_2, \star)$ be two semihypergroups and $f : H_1 \rightarrow H_2$ be a mapping. We say that f is a homomorphism of *type 5*, if for all $x, y \in H_1$, f is a good homomorphism and furthermore

$$(1)\quad f(x/y) = f(x/f^{-1}(f(y))),$$
$$(2)\quad f(y\backslash x) = f(f^{-1}(f(y))\backslash x);$$

type 6, if for all $x, y \in H_1$, f is a good homomorphism and furthermore

$$(3)\quad f(x/f^{-1}(f(y))) = f(x)/f(y),$$
$$(4)\quad f(f^{-1}(f(y))\backslash x) = f(y)\backslash f(x);$$

type 7, if for all $x, y \in H_1$, f is a good homomorphism and furthermore

$$(5)\quad f(x/y) = f(x)/f(y),$$
$$(6)\quad f(y\backslash x) = f(y)\backslash f(x).$$

Theorem 2.7.19. *If $f : H_1 \to H_2$ is a homomorphism of type 7, then f is a homomorphism of type 4.*

Proof. In general, if f is an inclusion homomorphism, then for every $x, y \in H_1$,

$$f^{-1}(f(x)) \circ f^{-1}(f(y)) \subseteq f^{-1}(f(x) \star f(y)).$$

Now, let f be a homomorphism of type 7. Suppose that $z \in f^{-1}(f(x)\star f(y))$. Then, $f(z) \in f(x)\star f(y)$, which implies that $f(y) \in f(x)\backslash f(z)$ and so $f(y) \in f(x\backslash z)$. Thus, there exists $y' \in x\backslash z$ such that $f(y) = f(y')$, consequently $z \in x \circ y' \subseteq f^{-1}(f(x)) \circ f^{-1}(f(y))$. Therefore, $f^{-1}(f(x)\star f(y)) \subseteq f^{-1}(f(x)) \circ f^{-1}(f(y))$. □

Theorem 2.7.20. *If $f : H_1 \to H_2$ is an onto homomorphism of type 4, then f is a homomorphism of type 6.*

Proof. We know that an onto homomorphism of type 4 is a good homomorphism. Suppose that $u \in f(z)/f(x)$. Then, there exists y such that $f(y) = u$ and so $f(z) \in f(y) \star f(x)$. Consequently, $z \in f^{-1}(f(y) \star f(x)) = f^{-1}(f(y)) \circ f^{-1}(f(x))$, it follows that there exist a, b such that $z \in a \circ b$, where $a \in f^{-1}(f(y))$ and $b \in f^{-1}(f(x))$. Hence, $f(y) = f(a)$, $f(x) = f(b)$ and so $u = f(a) \in f(z/b) \subseteq f(z/f^{-1}(f(x)))$. Therefore, $f(z)/f(x) \subseteq f(z/f^{-1}(f(x)))$. Note that the inverse inclusion is always true. Similarly, we can prove $f(f^{-1}(f(y))\backslash x) = f(y)\backslash f(x)$. □

Remark 13. Note that, in general, a homomorphism of type 4 is not good.

2.8 REGULAR AND STRONGLY REGULAR RELATIONS

Let $(H, \circ)$ be a semihypergroup and ρ be an equivalence relation on H. If A and B are non-empty subsets of H, then

$$A\overline{\rho}B \text{ means that } \forall a \in A, \exists b \in B \text{ such that } a\rho b \text{ and } \forall b' \in B, \exists a' \in A \text{ such that } a'\rho b';$$

$$A\overline{\overline{\rho}}B \text{ means that } \forall a \in A, \forall b \in B, \text{ we have } a\rho b.$$

Definition 2.8.1. The equivalence relation ρ is called

(1) *regular on the right (on the left)* if for all x of H, from $a\rho b$, it follows that $(a \circ x)\overline{\rho}(b \circ x)$ $((x \circ a)\overline{\rho}(x \circ b)$ respectively);

(2) *strongly regular on the right (on the left)* if for all x of H, from $a\rho b$, it follows that $(a \circ x)\overline{\overline{\rho}}(b \circ x)$ $((x \circ a)\overline{\overline{\rho}}(x \circ b)$ respectively);

(3) ρ is called *regular (strongly regular)* if it is regular (strongly regular) on the right and on the left.

Theorem 2.8.2. *Let $(H, \circ)$ be a semihypergroup and ρ be an equivalence relation on H.*

(1) *If ρ is regular, then H/ρ is a semihypergroup, with respect to the following hyperoperation: $\overline{x} \otimes \overline{y} = \{\overline{z} \mid z \in x \circ y\}$;*

(2) *If the above hyperoperation is well-defined on H/ρ, then ρ is regular.*

Proof

(1) First, we check that the hyperoperation $\otimes$ is well-defined on H/ρ. Consider $\overline{x} = \overline{x_1}$ and $\overline{y} = \overline{y_1}$. We check that $\overline{x} \otimes \overline{y} = \overline{x_1} \otimes \overline{y_1}$. We have $x\rho x_1$ and $y\rho y_1$. Since ρ is regular, it follows that $(x \circ y)\overline{\rho}(x_1 \circ y)$, $(x_1 \circ y)\overline{\rho}(x_1 \circ y_1)$ whence $(x \circ y)\overline{\rho}(x_1 \circ y_1)$. Hence, for all $z \in x \circ y$, there exists $z_1 \in x_1 \circ y_1$ such that $z\rho z_1$, which means that $\overline{z} = \overline{z_1}$. It follows that $\overline{x} \otimes \overline{y} \subseteq \overline{x_1} \otimes \overline{y_1}$ and similarly we obtain the converse inclusion. Now, we check the associativity of $\otimes$. Let $\overline{x}, \overline{y}, \overline{z}$ be arbitrary elements in H/ρ and $\overline{u} \in (\overline{x} \otimes \overline{y}) \otimes \overline{z}$. This means that there exists $\overline{v} \in \overline{x} \otimes \overline{y}$ such that $\overline{u} \in \overline{v} \otimes \overline{z}$. In other words, there exist $v_1 \in x \circ y$ and $u_1 \in v \circ z$, such that $v\rho v_1$ and $u\rho u_1$. Since ρ is regular, it follows that there exists $u_2 \in v_1 \circ z \subseteq x \circ (y \circ z)$ such that $u_1\rho u_2$. From here, we obtain that there exists $u_3 \in y \circ z$ such that $u_2 \in x \circ u_3$. We have $\overline{u} = \overline{u_1} = \overline{u_2} \in \overline{x} \otimes \overline{u_3} \subseteq \overline{x} \otimes (\overline{y} \otimes \overline{z})$. It follows that $(\overline{x} \otimes \overline{y}) \otimes \overline{z} \subseteq \overline{x} \otimes (\overline{y} \otimes \overline{z})$. Similarly, we obtain the converse inclusion.

(2) Let $a\rho b$ and x be an arbitrary element of H. If $u \in a \circ x$, then $\overline{u} \in \overline{a} \otimes \overline{x} = \overline{b} \otimes \overline{x} = \{\overline{v} | v \in b \circ x\}$. Hence, there exists $v \in b \circ x$ such that

$u\rho v$, whence $(a \circ x)\overline{\rho}(b \circ x)$. Similarly we obtain that ρ is regular on the left.

□

Corollary 2.8.3. *If $(H, \circ)$ is a hypergroup and ρ is an equivalence relation on H, then ρ is regular if and only if $(H/\rho, \otimes)$ is a hypergroup.*

Proof. If H is a hypergroup, then for all x of H we have $H \circ x = x \circ H = H$, whence we obtain $H/\rho \otimes \overline{x} = \overline{x} \otimes H/\rho = H/\rho$. According to Theorem 2.8.2, it follows that $(H/\rho, \otimes)$ is a hypergroup. □

Notice that if ρ is regular on a semihypergroup $(H, \circ)$, then the canonical projection $\pi : H \to H/\rho$ is a good epimorphism. Indeed, for all x, y of H and $\overline{z} \in \pi(x \circ y)$, there exists $z' \in x \circ y$ such that $\overline{z} = \overline{z'}$. We have $\overline{z} = \overline{z'} \in \overline{x} \otimes \overline{y} = \pi(x) \otimes \pi(y)$. Conversely, if $\overline{z} \in \pi(x) \otimes \pi(y) = \overline{x} \otimes \overline{y}$, then there exists $z_1 \in x \circ y$ such that $\overline{z} = \overline{z_1} \in \pi(x \circ y)$.

Theorem 2.8.4. *If $(H, \circ)$ and $(K, \star)$ are semihypergroups and $f : H \to K$ is a good homomorphism, then the equivalence ρ^f associated with f, that is $x\rho^f y \Leftrightarrow f(x) = f(y)$, is regular and $\varphi : f(H) \to H/\rho^f$, defined by $\varphi(f(x)) = \overline{x}$, is an isomorphism.*

Proof. Let $h_1 \rho^f h_2$ and a be an arbitrary element of H. If $u \in h_1 \circ a$, then

$$f(u) \in f(h_1 \circ a) = f(h_1) \star f(a) = f(h_2) \star f(a) = f(h_2 \circ a).$$

Hence, there exists $v \in h_2 \circ a$ such that $f(u) = f(v)$, which means that $u\rho^f v$. Hence, ρ^f is regular on the right. Similarly, it can be shown that ρ^f is regular on the left. On the other hand, for all $f(x), f(y)$ of $f(H)$, we have

$$\varphi(f(x) \star f(y)) = \varphi(f(x \circ y)) = \{\overline{z} \mid z \in x \circ y\} = \overline{x} \otimes \overline{y} = \varphi(f(x)) \otimes \varphi(f(y)).$$

Moreover, if $\varphi(f(x)) = \varphi(f(y))$, then $x\rho^f y$, so φ is injective and clearly, it is also surjective. Finally, for all $\overline{x}, \overline{y}$ of H/ρ^f we have

$$\begin{aligned}\varphi^{-1}(\overline{x} \otimes \overline{y}) &= \varphi^{-1}(\{\overline{z} \mid z \in x \circ y\}) = \{f(z) \mid z \in x \circ y\} \\ &= f(x \circ y) = f(x) \star f(y) = \varphi^{-1}(\overline{x}) \star \varphi^{-1}(\overline{y}).\end{aligned}$$

Therefore, φ is an isomorphism. □

Theorem 2.8.5. *Let $(H, \circ)$ be a semihypergroup and ρ be an equivalence relation on H.*

(1) *If ρ is strongly regular, then H/ρ is a semigroup, with respect to the following operation: $\overline{x} \otimes \overline{y} = \overline{z}$, for all $z \in x \circ y$;*

(2) *If the above operation is well-defined on H/ρ, then ρ is strongly regular.*

Proof. (1) For all x, y of H, we have $(x \circ y)\overline{\overline{\rho}}(x \circ y)$. Hence, $\overline{x} \otimes \overline{y} = \{\overline{z} \mid z \in x \circ y\} = \{\overline{z}\}$, which means that $\overline{x} \otimes \overline{y}$ has exactly one element. Therefore, $(H/\rho, \otimes)$ is a semigroup.

(2) If $a\rho b$ and x is an arbitrary element of H, we check that $(a\circ x)\overline{\overline{\rho}}(b\circ x)$. Indeed, for all $u \in a\circ x$ and all $v \in b\circ x$ we have $\overline{u} = \overline{a}\otimes\overline{x} = \overline{b}\otimes\overline{x} = \overline{v}$, which means that $u\rho v$. Hence, ρ is strongly regular on the right and similarly, it can be shown that it is strongly regular on the left. □

Theorem 2.8.6. *If $(H, \circ)$ is a semihypergroup, $(S, \star)$ is a semigroup, and $f : H \to S$ is a good homomorphism, then the equivalence ρ^f associated with f is strongly regular.*

Proof. Let $a\rho^f b$, $x \in H$ and $u \in a \circ x$. It follows that

$$f(u) = f(a) \star f(x) = f(b) \star f(x) = f(b \circ x).$$

Hence, for all $v \in b \circ x$, we have $f(u) = f(v)$, which means that $u\rho^f v$. Hence, ρ^f is strongly regular on the right and similarly, it is strongly regular on the left. □

Let α be a good homomorphism from a semihypergroup $(H, \circ)$ into a semihypergroup $(H', \star)$. The relation $\alpha^{-1} \circ \alpha$ is an equivalence relation ρ on H. This means that

$$a\rho b \Leftrightarrow \alpha(a) = \alpha(b).$$

The relation ρ is called the *kernel* of α. The natural mapping associated with ρ is $\varphi : H \to H/ker\alpha$, where $\varphi(a) = \rho(a)$. The mapping $\psi : H/\rho \to H'$, where $\psi(\rho(a)) = \alpha(a)$ is then the unique good homomorphism that makes the following diagram commute.

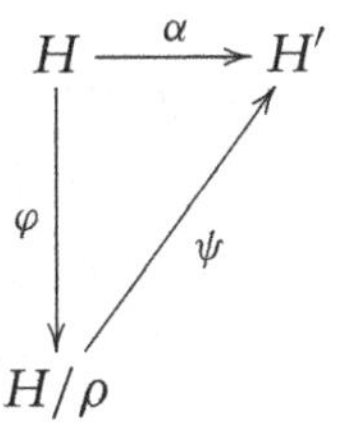

Theorem 2.8.7. *Let α_1 and α_2 be good homomorphisms of a semihypergroup $(H, \star)$ onto semihypergroups $(H_1, \star_1)$ and $(H_2, \star_2)$ respectively, such that $\alpha_1^{-1} \circ \alpha_1 \subseteq \alpha_2^{-1} \circ \alpha_2$. Then, a unique good homomorphism $\theta : H_1 \to H_2$ such that $\theta \circ \alpha_1 = \alpha_2$, exists. This means that the following diagram commutes.*

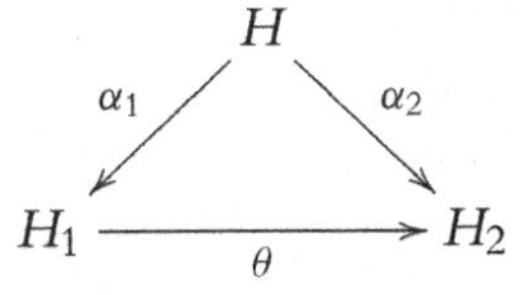

Proof. Let $a_1 \in H_1$ and a be an element of H such that $\alpha_1(a) = a_1$. Define $\theta(a_1) = \alpha_2(a)$. This is single valued, because or indeed if $\alpha_1(b) = a_1$ (b in H), we have $(a, b) \in \alpha_1^{-1} o \alpha_1 \subseteq \alpha_2^{-1} o \alpha_2$, so that $\alpha_2(a) = \alpha_2(b)$. It is clear that $\theta \circ \alpha_1 = \alpha_2$ and the assertion that θ is a good homomorphism follows from this:

$$\begin{aligned}\theta(\alpha_1(a) \star_1 \alpha_1(b)) &= \theta(\alpha_1(a \star b)) = \alpha_2(a \star b)\\ &= \alpha_2(a) \star_2 \alpha_2(b) = \theta(\alpha_1(a)) \star_2 \theta(\alpha_1(b)).\end{aligned}$$

The uniqueness of θ is evident, because or indeed, if θ is to satisfy $\theta \circ \alpha_1 = \alpha_2$ we are compelled to define θ as above. □

Theorem 2.8.8. *Let $(H, \star)$ and $(H', \star')$ be two semihypergroups and $\alpha : H \to H'$ be a good homomorphism. Then, $\rho = ker\alpha$ is a regular relation and there exists a good homomorphism $f : H/\rho \to H'$ such that $f \circ \varphi = \alpha$.*

Proof. Suppose that $a\rho b$. Then, for every $c \in H$, we have

$$\alpha(a \star c) = \alpha(a) \star' \alpha(c) = \alpha(b) \star' \alpha(c) = \alpha(b \star c).$$

Thus, for every $x \in a \star c$, there exists $y \in b \star c$ such that $\alpha(x) = \alpha(y)$ and so $x\rho y$. Therefore, ρ is a regular relation on H.

Now, let $\rho(a) \in H/\rho$ and define $f(\rho(a)) = \alpha(a)$. We note that if $b \in \rho(a)$, then $\alpha(a) = \alpha(b)$. So, f is well-defined. Moreover, f is a good homomorphism because if $\rho(a),\ \rho(b) \in H/\rho$, then

$$\begin{aligned}f(\rho(a) \odot \rho(b)) &= f(\{\rho(z) \mid z \in a \star b\})\\ &= \{\alpha(z) \mid z \in a \star b\}\\ &= \alpha(a \star b)\\ &= \alpha(a) \star' \alpha(b)\\ &= f(\rho(a)) \star' f(\rho(b)).\end{aligned}$$

Therefore, f is a good homomorphism. □

Theorem 2.8.9. *If ρ_1 and ρ_2 are strongly regular relations on a semihypergroup H such that $\rho_1 \subseteq \rho_2$, then there exists a good homomorphism from H/ρ_1 onto H/ρ_2.*

Proof. Let $\pi_1 : H \to H/\rho_1$ and $\pi_2 : H \to H/\rho_2$ be the canonical homomorphisms. Since $\rho_1 = \pi_1^{-1} o \pi_1$ and $\rho_2 = \pi_2^{-1} o \pi_2$, the hypotheses of Theorem 2.8.7 are satisfied and we conclude that there is a good homomorphism θ of H/ρ_1 onto H/ρ_2. □

Lemma 2.8.10. *Let I be a hyperideal of a semihypergroup $(H, \star)$. We consider the Rees relation on H as follows:*

$$a\rho b \Leftrightarrow a = b \text{ or } (a \in I \text{ and } b \in I).$$

Then, ρ is a strongly regular relation on H.

Proof. Obviously, ρ is an equivalence relation. Now, let $x\rho y$ and $a \in H$. If both x, y belong to I, then $x \star a \subseteq I$ and $y \star a \subseteq I$. Therefore, for every $u \in x \star a$ and $v \in y \star a$, we have $u\rho v$. If $x = y$, then $x \star a = y \star a$ and so for every $u \in x \star a$ we have $(u, u) \in \rho$. Therefore, ρ is a strongly regular relation on H. □

Theorem 2.8.11. *Let A be a hyperideal and B a subsemihypergroup of a semihypergroup $(H, \star)$. Then, $A \cap B$ is a hyperideal of B, $A \cup B$ is a subsemihypergroup of H, and there is an inclusion homomorphism from $\frac{B}{A\cap B}$ onto $\frac{A\cup B}{A}$.*

Proof. Since

$$(A \cup B)^2 = A^2 \cup A \star B \cup B \star A \cup B^2 \in \mathcal{P}^*(A \cup B),$$

it follows that $A \cup B$ is a subsemihypergroup of H. It is evident that $A \cap B$ is a hyperideal of B, and A is a hyperideal of $A \cup B$. Hence, the quotients $(A \cup B)/A$ and $B/A \cap B$ are defined. Now, if we place:

1. on B: $x\rho_1 y \Leftrightarrow x = y$ or $(x \in A \cap B$ and $y \in A \cap B)$,
2. on $A \cup B$: $x\rho_2 y \Leftrightarrow x = y$ or $(x \in A$ and $y \in A)$.

For every $x \in B$, we have $\rho_1(x) \subseteq \rho_2(x)$. Now, we consider the map $f : \frac{B}{A\cap B} \to \frac{A\cup B}{A}$ with $f(\rho_1(x)) = \rho_2(x)$. It is easy to see that f is well-defined. We show that f is an inclusion homomorphism. For every $x, y \in B$ we have

$$\begin{aligned} f(\rho_1(x) \odot \rho_1(y)) &= \{f(\rho_1(z)) | z \in \rho_1(x) \star \rho_1(y)\} \\ &= \{\rho_2(z) | z \in \rho_1(x) \star \rho_1(y)\} \\ &\subseteq \{\rho_2(z) | z \in \rho_2(x) \star \rho_2(y)\} \\ &= \rho_2(x) \otimes \rho_2(y). \end{aligned}$$

□

2.9 SIMPLE SEMIHYPERGROUPS

Jafarabadi, Sarmin, and Molaei introduced and analyzed the notion of simple and completely simple semihypergroups. In particular, they considered the quotient of semihypergroups and proved that if H is a simple semihypergroup and ρ a regular relation on H, then H/ρ is a simple semihypergroup. In this section, we study the properties of simple and completely simple semihypergroups. The main reference for this section is [87].

Definition 2.9.1. A semihypergroup $(H, \circ)$ without zero element is called *simple* if it has no proper hyperideals. A semihypergroup $(H, \circ)$ with zero element is called *0-simple* if it has the following conditions.

(1) $\{0\}$ and H are only its hyperideals;

(2) $H^2 \neq 0$.

Example 46. The semihypergroup defined in Example 21(6) is a simple semihypergroup.

Proposition 2.9.2. *Semihypergroup $(H, \circ)$ is 0-simple if and only if $H \circ a \circ H = H$, for all $a \in H$. It means for every $a, b \in H \setminus \{0\}$, there exist $x, y \in H$ such that $b \in x \circ a \circ y$.*

Proof. Suppose that H is a 0-simple semihypergroup. It is easy to see that H^2 is a hyperideal of H. By the definition, $H^2 \neq 0$. Thus, $H^2 = H$, and it follows that $H^3 = H$. Now, we consider an element $a \in H$ that is not a zero element. It is clear that $H \circ a \circ H$ is a hyperideal of H. In the case of $H \circ a \circ H = \{0\}$, the set $I = \{h \in H | H \circ h \circ H = \{0\}\}$ is a non-empty subset of H since $a \in I$. If x is an element of $H \circ I$, then there exist elements $h \in H$ and $i \in I$ such that $x \in h \circ i$ and hence $H \circ x \circ H \subseteq H \circ h \circ i \circ H \subseteq H \circ i \circ H = \{0\}$. This implies $H \circ x \circ H = \{0\}$ and $x \in I$. In a similar way, it can be shown that $I \circ H$ is also a subset of I. It follows that I is a hyperideal of H, and so $I = H$. Thus, for all $h \in H$, $H \circ h \circ H = \{0\}$, that is $H^3 = \{0\}$, which is a contradiction to $H^3 = H$. Therefore, $H \circ a \circ H = H$, for every $a \neq 0$ in H.

Conversely, assume that $H \circ a \circ H = H$, for all $a \neq 0$ in H. Then, H^2 is not equal to $\{0\}$ (if $H^2 = \{0\}$, then $H \circ a \circ H = \{0\}$, for all $a \neq 0$ in H). Now, suppose that a is an element of H that is not a zero element and I is a hyperideal of H containing a. Then, we have $H = H \circ a \circ H \subseteq H \circ I \circ H \subseteq I$. This implies that $H = I$. Thus, H is 0-simple. □

Proposition 2.9.2 leads to the following important corollary that can be used for a characterization of simple semihypergroups.

Corollary 2.9.3. *A semihypergroup $(H, \circ)$ is simple if and only if for all $a \in H$, $H \circ a \circ H = H$.*

Example 47. Apart from isomorphisms, all simple semihypergroups of size 2 are the following.

$\circ_1$	a	b
a	a	a
b	b	b

$\circ_2$	a	b
a	a	b
b	a	b

$\circ_3$	a	b
a	a	b
b	b	a

$\circ_4$	a	b
a	a	$\{a, b\}$
b	a	b

$\circ_5$	a	b
a	a	$\{a, b\}$
b	a	$\{a, b\}$

$\circ_6$	a	b
a	a	a
b	$\{a, b\}$	b

$\circ_7$	a	b
a	a	a
b	$\{a, b\}$	$\{a, b\}$

Proposition 2.9.4. *Every hypergroup is a simple semihypergroup.*

Proof. If $(H, \circ)$ is a hypergroup, then for all $a \in H$, $a \circ H = H \circ a = H$. Hence, we obtain $H = a \circ H \subseteq H \circ H \subseteq H$ and so $H \circ H = H$. On the other hand, $a \circ H = H$, so $H \circ a \circ H = H \circ H = H$, ie, H is simple. □

Theorem 2.9.5. *Let $(H, \circ)$ and $(H', \star)$ be two simple semihypergroups. Then, the product $H \times H'$ with the hyperoperation defined in Example 21(9) is simple semihypergroup.*

Proof. It is straightforward. □

Proposition 2.9.6. *Let $(H, \circ)$ be a simple semihypergroup and ρ be a regular relation on H. Then, H/ρ is a simple semihypergroup.*

Proof. According to Theorem 2.8.2, H/ρ is a semihypergroup. Suppose that $\bar{a}$ and $\bar{b}$ are two arbitrary elements of H/ρ. Since H is simple, it follows that there exist elements x and y in H such that $b \in x \circ a \circ y$. Thus, $\bar{b} \in \bar{x} \otimes \bar{a} \otimes \bar{y}$. Thus, we conclude that H/ρ is simple. □

Example 48. Consider the set of integer numbers $\mathbb{Z}$ with the hyperoperation $i \circ j = \{i, j\}$. By Proposition 2.9.4, this hypergroup is a simple semihypergroup. Let the equivalence relation ρ be the congruence modulo 2, that is a regular relation. Then, $\mathbb{Z}/\rho$ with respect to the following hyperoperation is a simple semihypergroup,

$$\bar{i} \otimes \bar{j} = \{\bar{k} | k \in i \circ j = \{i, j\}\}.$$

The following theorem is an approach to a generalization of Rees theorem in semigroup theory. By using this theorem, new simple semihypergroups can also be constructed.

Theorem 2.9.7. *Let* $(H, \circ)$ *be a regular hypergroup and* I, Λ *be non-empty sets. Let* $P = (p_{ij})$ *be a* $\Lambda \times I$ *regular matrix (it has no row or column that consists entirely of zeros) with entries from* H. *Then* $S = I \times H \times \Lambda$ *(Rees Matrix Semihypergroup) with respect to the following hyperoperation is a simple semihypergroup,*

$$(i, a, \lambda) \star (j, b, \mu) = \{(i, t, \mu) | t \in a \circ p_{\lambda_j} \circ b\}.$$

Proof. In a direct verification, the associativity of the hyperoperation can be proved. Let (i, a, λ), (j, b, μ), and (k, c, ν) be arbitrary elements of S, and z is an element of the following set

$$(i, a, \lambda) \star \big((j, b, \mu) \star (k, c, \nu)\big) = \bigcup_{t \in b \circ p_{\mu k} \circ c} \{(i, x, \nu) | x \in a \circ p_{\lambda j} \circ t\}.$$

There exists $t' \in b \circ p_{\mu k} \circ c$ and $x' \in a \circ p_{\lambda_j} \circ t'$ such that $z = (i, x', \nu)$. This means there exists $v \in p_{\mu k} \circ c$ and $u \in a \circ p_{\lambda j}$ such that $t' \in b \circ v$ and $x' \in u \circ t'$, and so $x' \in u \circ b \circ v$. It follows that there exists $s \in a \circ p_{\lambda j} \circ b$ such that $x' \in s \circ p_{\mu k} \circ c$. Thus,

$$z = (i, x', \nu) \in \bigcup_{s \in a \circ p_{\lambda j} \circ b} \{(i, y, \nu) | y \in s \circ p_{\mu k} \circ c\} = \big((i, a, \lambda) \star (j, b, \mu)\big) \star (k, c, \nu).$$

In a similar way, it can be shown that every element of $\big((i, a, \lambda) \star (j, b, \mu)\big) \star (k, c, \nu)$ is an element of the set $(i, a, \lambda) \star \big((j, b, \mu) \star (k, c, \nu)\big)$. Therefore, S is a semihypergroup.

In order to verify that S is simple, suppose that (i, a, λ) and (j, b, μ) are two elements of S that are not zero elements. Since H is a hypergroup, it follows that there exist elements x and y in H such that $b \in x \circ a \circ y$, and due to the regularity of H, there exists an identity element e in H such that $b \in x \circ e \circ a \circ e \circ y$, and so $b \in x \circ p_{\nu i}^{-1} \circ p_{\nu i} \circ a \circ p_{\lambda k} \circ p_{\lambda k}^{-1} \circ y$ (by the regularity of matrix P, the elements $\nu \in \Lambda$ and $k \in I$ can be chosen such that $p_{\lambda k}$ and $p_{\nu i}$ are not zero element). It follows that there exists $t \in x \circ p_{\nu i}^{-1} \circ p_{\nu i} \circ a$ such that $b \in t \circ p_{\lambda k} \circ p_{\lambda k}^{-1} \circ y$. Hence,

$$\begin{aligned}(j, b, \mu) &\in \{(j, e, \nu) | e \in x \circ p_{\nu i}^{-1}\} \star (i, a, \lambda) \star \{(k, f, \mu) | f \in p_{\lambda k}^{-1} \circ y\} \\ &= \bigcup_{t \in x \circ p_{\nu i}^{-1} \circ p_{\nu i} \circ a} \{(j, s, \mu) \, | s \in t \circ p_{\lambda k} \circ p_{\lambda k}^{-1} \circ y\}.\end{aligned}$$

It means that there exists $e' \in x \circ p_{\nu i}^{-1}$ and $f' \in p_{\lambda k}^{-1} \circ y$ such that

$$(j, b, \mu) \in (j, e', \nu) \star (i, a, \lambda) \star (k, f', \mu).$$

Thus, S is simple. □

Definition 2.9.8. An element e in a semihypergroup $(H, \circ)$ is called an *idempotent* if $e \in e^2$.

In the set of all scalar idempotent elements of semihypergroup $(H, \circ)$, we can define an order $e \leq f$ if and only if $e = e \circ f = f \circ e$. It is easy to show that this relation is a partial order relation.

Definition 2.9.9. A scalar idempotent e in the set of all scalar idempotent elements of semihypergroup $(H, \circ)$ is called *primitive scalar idempotent* (or just *primitive*) if it is minimal within the set of all non-zero scalar idempotent elements of H. Thus, a primitive scalar idempotent has the following property: If $0 \neq f = e \circ f = f \circ e$, then $f = e$.

Definition 2.9.10. A semihypergroup $(H, \circ)$ is called a *completely simple semihypergroup* if it is simple and has a primitive idempotent.

Example 49. Consider semihypergroup $H = \{p, q, r, t\}$ with respect to the following Cayley table.

$\circ$	p	q	r	t
p	p	q	r	t
q	q	$\{p, r\}$	$\{q, r\}$	t
r	r	$\{q, r\}$	$\{p, q\}$	t
t	t	t	t	H

In this semihypergroup, p and t are idempotent elements. The idempotent p is the only scalar idempotent, that is, a primitive idempotent. This semihypergroup is clearly simple ($H \circ a \circ H = H$, for all $a \in H$). So, it is a completely simple semihypergroup.

Example 50. The semihypergroup of Example 46 is a completely simple semihypergroup. In fact, every element of this semihypergroup is a primitive idempotent.

Lemma 2.9.11. *Every scalar idempotent of a regular hypergroup is a scalar identity.*

Proof. Let $(H, \circ)$ be a regular hypergroup and a be a scalar idempotent of H. Then, $a^2 = a$, so $a^{-1} \circ a^2 = a^{-1} \circ a$ (due to the regularity of H, a^{-1} as an inverse of a exists). It means there exists an identity element such as $e \in H$ such that $e \circ a = e$ (a is scalar). On the other hand, e is the identity. Thus, we have $e \circ a = a \circ e = a$, and so $a = e$. □

Proposition 2.9.12. *Every regular hypergroup is a completely simple semihypergroup.*

Proof. Let $(H, \circ)$ be a regular hypergroup. Assume that e and f are two scalar idempotent elements of H, and $f = e \circ f = f \circ e$. Then, f is an identity element of H. Thus, $e = e \circ f = f \circ e$, and so $e = f$. It means every scalar idempotent of H is primitive and this implies that H is a completely simple semihypergroup. □

To construct new completely simple semihypergroups by using the quotient semihypergroups, the following theorem will be used.

Theorem 2.9.13. *Let* $(H, \circ)$ *be a regular hypergroup and* ρ *be a regular relation on* H. *Then,* $(H/\rho, \otimes)$ *is a regular hypergroup.*

Proof. According to Corollary 2.8.3, since H is a hypergroup, it follows that H/ρ is also a hypergroup. Suppose that e is an identity element of H. Then, $x \in x \circ e \cap e \circ x$, for all $x \in H$, so $\overline{x} \in \overline{x \circ e} \cap \overline{e \circ x}$. Canonical projection $\pi : H \to H/\rho$ is a good epimorphism, thus $\overline{x} \in \overline{x} \otimes \overline{e} \cap \overline{e} \otimes \overline{x}$. It means $\overline{e}$ is an identity element of H/ρ. On the other hand, if $\overline{x}$ is an arbitrary element of H/ρ, then there exists $x \in H$ such that $\pi(x) = \overline{x}$. By the regularity of H, there exist elements such as e (identity) and x' in H such that $e \in x \circ x' \cap x' \circ x$ and so $\overline{e} \in \overline{x \circ x'} \cap \overline{x' \circ x} = \overline{x} \otimes \overline{x'} \cap \overline{x'} \otimes \overline{x}$. It means that $\overline{x'}$ is an inverse for $\overline{x}$. □

Example 51. Consider the set of integers $(\mathbb{Z}, +)$ as a regular hypergroup. Let the equivalence relation ρ be the congruence modulo 2, that is, a regular relation. Then, it is easy to show that $\mathbb{Z}/\rho = \{\overline{0}, \overline{1}\}$ with respect to the following hyperoperation,

$$\overline{i} \otimes \overline{j} = \{\overline{k} | k = i + j\}, \text{ for all } \overline{i}, \overline{j} \in \mathbb{Z}/\rho,$$

is a regular semihypergroup.

Corollary 2.9.14. *Let* $(H, \circ)$ *be a regular hypergroup and* ρ *be a regular relation on* H. *Then,* $(H/\rho, \otimes)$ *is a completely simple semihypergroup.*

Proof. The proof follows from Theorem 2.9.13 and Proposition 2.9.12. □

Proposition 2.9.15. *Every finite hypergroup with at least one scalar element is a completely simple semihypergroup.*

Proof. Let H be a finite hypergroup with at least one scalar element. Then, there exists an identity element of H such as e such that the set of all scalar elements of H is a group with the identity e. It is clear that e is a scalar idempotent. It now follows that the primitive scalar idempotent exists. Consider a descending chain $e_1 \geq e_2 \geq e_3 \geq \cdots$ of scalar idempotent elements in H. This chain has to be finite because H is finite. It means H has primitive scalar idempotent, and so it is a completely simple semihypergroup. □

Theorem 2.9.16. *Let* $(H, \circ)$ *and* $(H', \star)$ *be two completely simple semihypergroups. Then, the product* $H \times H'$ *is a completely simple semihypergroup.*

Proof. It is straightforward. □

Theorem 2.9.17. *Let $(H, \circ)$ be a regular hypergroup, and I, Λ be non-empty sets. Let $P = (p_{ij})$ be a $\Lambda \times I$ regular matrix (it has no row or column that consists entirely of zeros) with entries from H. Then, $S = I \times H \times \Lambda$ (Rees Matrix Semihypergroup) with respect to the hyperoperation*

$$(i, a, \lambda) \star (j, b, \mu) = \{(i, t, \mu) | t \in a \circ p_{\lambda j} \circ b\}$$

is a completely simple semihypergroup.

Proof. According to Theorem 2.9.7, S is simple. It is easy to show that every scalar idempotent of S is primitive idempotent. Therefore, $(S, \star)$ is a completely simple semihypergroup. □

Theorem 2.9.18. *Let $f : H \to H'$ be a good homomorphism from a regular semihypergroup H into a semihypergroup H'. Then, Imf is a regular semihypergroup.*

Proof. It is straightforward. □

Corollary 2.9.19. *Let $(H, \circ)$ be a regular semihypergroup and ρ be a regular relation on H. Then, $(H/\rho, \otimes)$ is a regular semihypergroup.*

Proof. By Theorem 2.9.18, the proof is clear. □

2.10 CYCLIC SEMIHYPERGROUPS

Cyclic semihypergroups have been studied by Desalvo and Freni [52], Vougiouklis [93], and Leoreanu [63]. Cyclic semihypergroups are important not only in the sphere of finitely generated semihypergroups, but also for interesting combinatorial implications.

Let $(H, \circ)$ be a semihypergroup and P be a non-empty subset of H. We say that P is a *cyclic part* of H if there exists $x \in P$ such that

$$\forall a \in P, \ \exists n \in \mathbb{N} : a \in x^n.$$

The element x is called the *generator* of P.

Definition 2.10.1. We say that H is a *cyclic semihypergroup* if H is a cyclic part.

So, a semihypergroup $(H, \circ)$ is cyclic if

$$H = x^1 \cup x^2 \cup \cdots \cup x^n \cup \cdots, \text{ for some } x \in H. \tag{2.5}$$

If there exists an integer $n > 0$, the minimum one with the following property

$$H = x^1 \cup x^2 \cup \cdots \cup x^n, \tag{2.6}$$

then we call H a cyclic semihypergroup with *finite period* and we call x the *generator of H with period n.* If there is no number n for which Eq. (2.6) is valid, but Eq. (2.5) is valid, then we say that H has *infinite period* for x.

Example 52

(1) The semihypergroup $H = \{a,\ b,\ c\}$ with the following hyperoperation is a cyclic semihypergroup.

$\circ$	a	b	c
a	b	$\{b,c\}$	$\{b,c\}$
a	$\{b,c\}$	$\{b,c\}$	$\{b,c\}$
a	$\{b,c\}$	$\{b,c\}$	$\{b,c\}$

Indeed, $a \in a^1$, $b \in a^2$, $c \in a^3$.

(2) The semihypergroup (hypergroup) $H = \{a,\ b,\ c,\ d\}$ with the following hyperoperation is a cyclic semihypergroup (hypergroup).

$\circ$	a	b	c	d
a	b	$\{a,c,d\}$	b	b
b	$\{a,c,d\}$	b	$\{a,c,d\}$	$\{a,c,d\}$
c	b	$\{a,c,d\}$	b	b
d	b	$\{a,c,d\}$	b	b

Indeed, $a \in a^1$, $b \in a^2$, $c \in a^3$, $d \in a^3$.

Proposition 2.10.2. *Let $(H, \circ)$ be a semihypergroup and A, B be cyclic subsemihypergroups of H with generators a, b, respectively.*

(1) *If $a \in A \cap B$, then $A \subseteq B$.*

(2) *If $\{a,\ b\} \subseteq A \cap B$, then $A = B$.*

Proof. (1) For every $x \in A$, there exists $m \in \mathbb{N}$ such that $x \in a^m$. Since $a \in B$, it follows that $a^m \subseteq B$. This implies that $x \in B$. Hence, $A \subseteq B$.

(2) This follows from (1). □

If $(H, \circ)$ is a semihypergroup, it is known that the intersection $\bigcap_{i \in I} S_i$ of a family $\{S_i\}_{i \in I}$ of subsemihypergroups of H, if it is non-empty, is a subsemihypergroup of H. For every non-empty subset $A \subseteq H$ there is at least one subsemihypergroup of H containing A, ie, H itself. Hence, the intersection of all subsemihypergroups of H containing A is a subsemihypergroup. We denote it by $\langle A \rangle$, and note that it is defined by the following two properties.

(1) $A \subseteq \langle A \rangle$;
(2) If S is a subsemihypergroup of H and $A \subseteq S$, then $\langle A \rangle \subseteq S$. Furthermore, $\langle A \rangle$ is characterized as the algebraic closure of A under the hyperproduct in H, namely we have $\langle A \rangle = \bigcup_{n \geq 1} A^n$. Moreover, if H is finite, the set $\{r \in \mathbb{N} | \bigcup_{k=1}^{r} A^k = \bigcup_{k=1}^{r+1} A^k\}$ has minimum $m \leq |H|$, otherwise

$$A \subset \bigcup_{k=1}^{2} A^k \subset \bigcup_{k=1}^{3} A^k \subset \cdots \subset \bigcup_{k=1}^{|H|} A^k \subset \bigcup_{k=1}^{|H|+1} A^k,$$

where all inclusions are proper and so $|\langle A \rangle| > |H|$, a contradiction. Then, we have

$$\langle A \rangle = \bigcup_{k=1}^{m} A^k = \bigcup_{k=1}^{m+1} A^k = \cdots = \bigcup_{k=1}^{|H|} A^k.$$

For the case $A = \{x\}$, a singleton set, we refer to $\langle x \rangle = \bigcup_{n \geq 1} x^n$ as the *cyclic subsemihypergroup* of H generated by the element x.

Remark 14. If $H = \langle x \rangle$, then H is a cyclic semihypergroup generated by x.

Definition 2.10.3. If $(H, \circ)$ is a semihypergroup and $P \subseteq H$ is a hyperproduct of elements in H, then the subsemihypergroup $\langle P \rangle$ is said to be *hypercyclic generated by* P. The semihypergroup $(H, \circ)$ is said to be *hypercyclic* if there exists a hyperproduct P of elements in H such that $H = \langle P \rangle$.

CHAPTER 3

Ordered Semihypergroups

3.1 BASIC DEFINITIONS AND EXAMPLES

The concept of ordering hypergroups was investigated by Chvalina [94] as a special class of hypergroups and studied by him and many others. In [95], Heidari and Davvaz studied a semihypergroup $(H, \circ)$ together with a binary relation $\leq$, where $\leq$ is a partial order relation that satisfies the monotone condition.

Definition 3.1.1. An *ordered semihypergroup* $(H, \circ, \leq)$ is a semihypergroup $(H, \circ)$ together with a partial order $\leq$ that is *compatible* with the hyperoperation, meaning that for any x, y, z in H,

$$x \leq y \Rightarrow z \circ x \leq z \circ y \quad \text{and} \quad x \circ z \leq y \circ z.$$

Here, $z \circ x \leq z \circ y$ means for any $a \in z \circ x$ there exists $b \in z \circ y$ such that $a \leq b$. The case $x \circ z \leq y \circ z$ is defined similarly.

Here are some examples of ordered semihypergroups. The main references are [96, 97].

Example 53. We have $(H, \circ, \leq)$ as an ordered semihypergroup where the hyperoperation and the order relation are defined by

$\circ$	a	b	c
a	a	$\{a, b\}$	$\{a, c\}$
b	a	$\{a, b\}$	$\{a, c\}$
c	a	$\{a, b\}$	c

$$\leq := \{(a, a), (b, b), (c, c), (a, b)\}.$$

The covering relation and the figure of H are given by: $\prec = \{(a, b)\}$

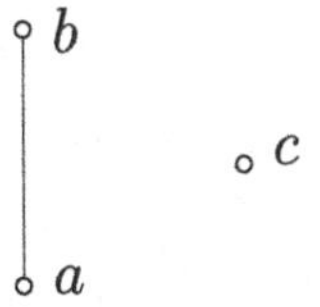

http://dx.doi.org/10.1016/B978-0-12-809815-8.00003-6

Example 54. We have $(H, \circ, \leq)$ as an ordered semihypergroup where the hyperoperation and the order relation are defined by

$\circ$	a	b	c	d
a	a	$\{a, b\}$	$\{a, c\}$	$\{a, d\}$
b	a	$\{a, b\}$	$\{a, c\}$	$\{a, d\}$
c	a	b	c	d
d	a	b	c	d

$$\leq := \{(a, a), (b, b), (c, c), (d, d), (a, b)\}.$$

The covering relation and the figure of H are given by: $\prec = \{(a, b)\}$

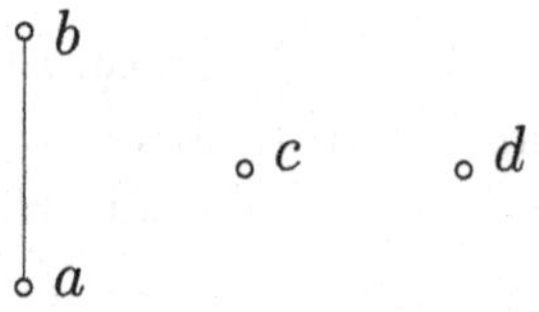

Example 55. Suppose that $H = \{x, y, z, r, s, t\}$. We consider the ordered semihypergroup $(H, \circ, \leq)$, where the hyperoperation $\circ$ is defined by the following table:

$\circ$	x	y	z	r	s	t
x	r	$\{r, s\}$	$\{r, t\}$	x	$\{x, y\}$	$\{x, z\}$
y	r	s	$\{r, t\}$	x	y	$\{x, z\}$
z	r	$\{r, s\}$	t	x	$\{x, y\}$	z
r	x	$\{x, y\}$	$\{x, z\}$	r	$\{r, s\}$	$\{r, t\}$
s	x	y	$\{x, z\}$	r	s	$\{r, t\}$
t	x	$\{x, y\}$	z	r	$\{r, s\}$	t

and the order $\leq$ is defined by

$$\leq := \{(x, x), (y, y), (z, z), (r, r), (s, s), (t, t), (s, r), (t, r), (y, x), (z, x)\}.$$

The covering relation and the figure of H are given by

$$\prec = \{(s, r), (t, r), (y, x), (z, x)\}.$$

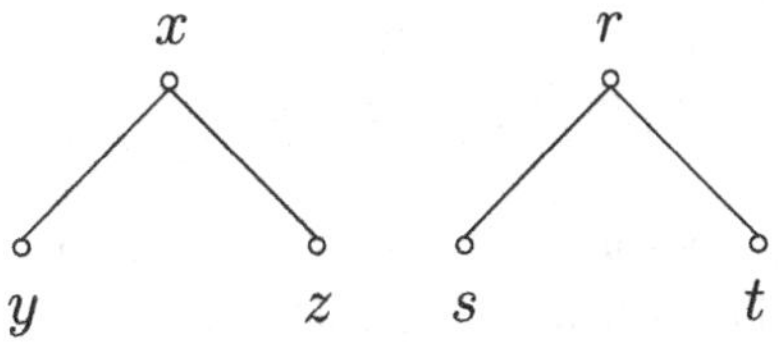

Definition 3.1.2. Let $(H, \circ, \leq_H)$ and $(T, \diamond, \leq_T)$ be two ordered semihypergroups. A mapping $f : H \to T$ is called a *homomorphism* if it satisfies the following conditions:

(1) $f(x \circ y) = f(x) \diamond f(y)$, for all $x, y \in H$;

(2) $x, y \in H$ and $x \leq_H y$ implies $f(x) \leq_T f(y)$.

Note that the concept of ordered semihypergroups is a generalization of the concept of ordered semigroups. Indeed, every ordered semigroup is an ordered semihypergroup.

For a non-empty subset A of an ordered semihypergroup $(H, \circ, \leq)$, we write

$$(A] = \{x \in H \mid x \leq a \text{ for some } a \in A\}.$$

The following is easy to see for non-empty subsets A, B of an ordered semihypergroup $(H, \circ, \leq)$:

(1) $A \subseteq (A]$;

(2) $A \subseteq B \Rightarrow (A] \subseteq (B]$;

(3) $(A] \circ (B] \subseteq (A \circ B]$;

(4) $((A] \circ (B]] = (A \circ B]$;

(5) $(A] \cup (B] = (A \cup B]$.

Definition 3.1.3. A non-empty subset A of an ordered semihypergroup $(H, \circ, \leq)$ is called a *right* (respectively, *left*) *hyperideal* of H if

(1) $A \circ H \subseteq A$ (respectively, $H \circ A \subseteq A$);

(2) for every $a \in H$, $b \in A$ and $a \leq b$ implies $a \in A$.

If A is both right hyperideal and left hyperideal of H, then A is called a *hyperideal (or two-side hyperideal)* of H.

Lemma 3.1.4. *Let $(H, \circ, \leq)$ be an ordered semihypergroup $a \in H$. Then $(a \circ H]$ is a right hyperideal and $(H \circ a]$ is a left hyperideal and $(H \circ a \circ H]$ is a hyperideal of H.*

Proof. Let $x \in H$ and $y \in (a \circ H]$. Then, there exist $z \in a \circ H$ and $h \in H$ such that $y \leq z \leq a \circ h$. We have

$$y \circ x \leq z \circ x \leq a \circ h \circ x \subseteq a \circ H.$$

Hence, $y \circ x \subseteq (a \circ H]$. If $x \leq y$, then $x \leq y \leq z$ and so, $x \in (a \circ H]$. Altogether, $(a \circ H]$ is a right hyperideal of H. The rest of the lemma can be proved similarly. □

Definition 3.1.5. A subsemihypergroup A of an ordered semihypergroup $(H, \circ, \leq)$ is called a *bi-hyperideal* of H if

(1) $A \circ H \circ A \subseteq A$; and

(2) for every $a \in H$, $b \in A$ and $a \leq b$ implies $a \in A$.

Example 56. Let $H = \{a, b, c, d, e, f\}$. We define a hyperoperation $\circ$ on H by the following table:

$\circ$	a	b	c	d	e	f
a	a	b	a	b	a	b
b	b	a	b	a	b	a
c	a	b	e	f	$\{a, c\}$	$\{b, d\}$
d	b	a	f	e	$\{b, d\}$	$\{a, c\}$
e	a	b	$\{a, c\}$	$\{b, d\}$	$\{a, e\}$	$\{b, f\}$
f	b	a	$\{b, d\}$	$\{a, c\}$	$\{b, f\}$	$\{a, e\}$

The order relation on H is defined as follows.

$$\leq := \{(a, a), (b, b), (c, c), (d, d), (e, e), (f, f), (a, c), (a, e), (b, f)\}.$$

We give the covering relation "$\prec$" and the figure of H:

$$\prec = \{(a, c), (a, e), (b, f)\}.$$

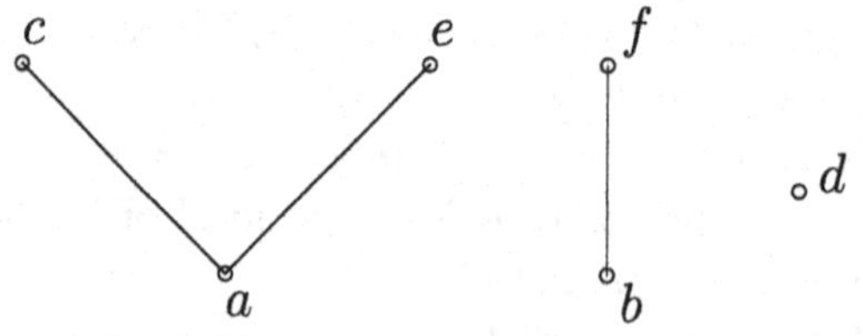

It is easily seen that $(H, \circ, \leq)$ is an ordered semihypergroup and $\{a, b\}$ is a hyperideal of H.

Example 57. Let $H = \{a, b, c, d, e\}$. We define hyperoperation $\circ$ on H by the following table:

$\circ$	a	b	c	d	e
a	a	a	$\{a, b, c\}$	a	$\{a, b, c\}$
b	a	a	$\{a, b, c\}$	a	$\{a, b, c\}$
c	a	a	$\{a, b, c\}$	a	$\{a, b, c\}$
d	$\{a, b, d\}$	$\{a, b, d\}$	H	$\{a, b, d\}$	H
e	$\{a, b, d\}$	$\{a, b, d\}$	H	$\{a, b, d\}$	H

We define order relation $\leq$ as follows:

$$\begin{aligned}\leq := \{&(a, a), (b, b), (c, c), (d, d), (e, e), (a, b), (a, c),\\ &(a, d), (a, e), (b, c), (b, d), (b, e), (c, e), (d, e)\}.\end{aligned}$$

We give the covering relation "$\prec$" and the figure of H:

$$\prec = \{(a, b), (a, c), (a, d), (a, e), (b, c), (b, d), (b, e), (c, e), (d, e)\}.$$

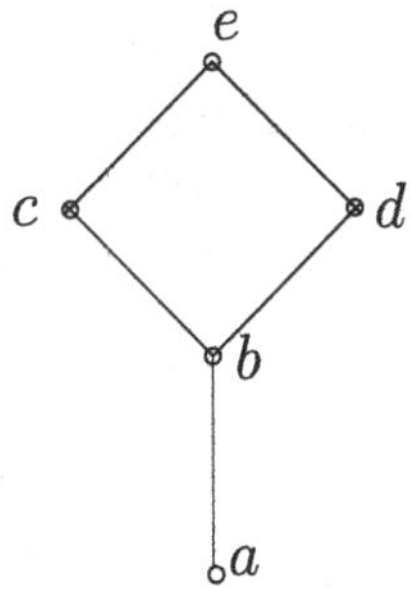

Now, $(H, \circ, \leq)$ is an ordered semihypergroup. It is easy to see that $\{a, b, c\}$ is a right hyperideal but not left hyperideal, $\{a, b, d\}$ is a left hyperideal but not right hyperideal, and $\{a\}$ is neither left hyperideal nor right hyperideal, but it is a bi-hyperideal of H.

Lemma 3.1.6. *Let $(H, \circ\ \leq)$ be an ordered semihypergroup and $\emptyset \neq A \subseteq H$. Then, the set $(A \cup A^2 \cup (A \circ H \circ A)]$ is the bi-hyperideal of H generated by A.*

Proof. Clearly, $(A \cup A^2 \cup (A \circ H \circ A)]$ is a subsemihypergroup of H. Let $h \in H$ and $a, b \in (A \cup A^2 \cup (A \circ H \circ A)]$. Then, there exist $x, y \in A \cup A^2 \cup (A \circ H \circ A)$ such that $a \leq x, b \leq y$. It is easy to see that $x \circ h \circ y \subseteq A \circ H \circ A$. So, $a \circ h \circ b \leq x \circ h \circ y \subseteq A \cup A^2 \cup (A \circ H \circ A)$. This implies $a \circ h \circ b \in (A \cup A^2 \cup (A \circ H \circ A)]$. Hence, $(A \cup A^2 \cup (A \circ H \circ A)] \circ H \circ (A \cup A^2 \cup (A \circ H \circ A)] \subseteq (A \cup A^2 \cup (A \circ H \circ A)]$. Now, we have that $(A \cup A^2 \cup (A \circ H \circ A)]$ is a bi-hyperideal of H, obviously, containing A.

Let B be a bi-hyperideal containing A. Let $z \in (A \cup A^2 \cup (A \circ H \circ A)]$. There is $w \in A \cup A^2 \cup (A \circ H \circ A)$ such that $z \leq w$. Since $A \subseteq B$ and B is a bi-hyperideal, it follows that $A^2 \subseteq B, A \circ H \circ A \subseteq B \circ H \circ B \subseteq B$. Thus, $z \leq w \in B$ and then $z \in B$. Altogether, $(A \cup A^2 \cup (A \circ H \circ A)]$ is the bi-hyperideal of H generated by A. □

Corollary 3.1.7. *Let $(H, \circ \leq)$ be an ordered semihypergroup and $a \in H$. Then, the set $(a \cup a^2 \cup (a \circ H \circ a)]$ is the bi-hyperideal of H generated by a.*

3.2 PRIME HYPERIDEALS OF THE CARTESIAN PRODUCT OF TWO ORDERED SEMIHYPERGROUPS

A hyperideal P of an ordered semihypergroup $(H, \circ, \leq)$ is said to be *prime* if $H \setminus P$ is a subsemihypergroup of H. Note that if a hyperideal P of H is prime, then $P \neq H$. In this section we accept the empty set to be a prime hyperideal. Changphas and Davvaz [96], similar to the method of Petrich [98], gave a necessary and sufficient condition of a subset of Cartesian product of two ordered semihypergroups to be a prime hyperideal.

Definition 3.2.1. Let $(H, \circ, \leq_H)$ and $(T, \diamond, \leq_T)$ be two ordered semihypergroups. Under the coordinatewise multiplication, ie,

$$(s_1, t_1) \star (s_2, t_2) = s_1 \circ s_2 \times t_1 \diamond t_2,$$

where $(s_1, t_1), (s_2, t_2) \in H \times T$, the Cartesian product $H \times T$ of H and T forms a semihypergroup. Define a partial order $\leq$ on $H \times T$ by

$$(s_1, t_1) \leq (s_2, t_2) \text{ if and only if } s_1 \leq_H s_2 \text{ and } t_1 \leq_T t_2,$$

where $(s_1, t_1), (s_2, t_2) \in H \times T$. Then, $(H \times T, \star, \leq)$ is an ordered semihypergroup.

Theorem 3.2.2. *Let $(H, \circ, \leq_H)$ and $(T, \diamond, \leq_T)$ be ordered semihypergroups. Then, a subset L of $H \times T$ is a prime hyperideal of $H \times T$ if and only if there exist a prime hyperideal I of H and a prime hyperideal J of T such that $L = (I \times T) \cup (H \times J)$.*

Proof. Assume that there exist a prime hyperideal I of H and a prime hyperideal J of T such that

$$L = (I \times T) \cup (H \times J).$$

If $I = \emptyset$ and $J = \emptyset$, then $L = \emptyset$; hence, L is a prime hyperideal of $H \times T$.

Suppose that $I \neq \emptyset$ or $J \neq \emptyset$. Then, $L \neq \emptyset$. We shall show that L is a prime hyperideal of $H \times T$. Let $(x, u) \in L$ and $(y, v) \in H \times T$. If $x \in I$, then $x \circ y \subseteq I$ and $y \circ x \subseteq I$; hence,

$$(x, u) \star (y, v) = x \circ y \times u \diamond v \subseteq I \times T$$

and

$$(y, v) \star (x, u) = y \circ x \times v \diamond u \subseteq I \times T.$$

Similarly, if $u \in J$, then $(x, u) \star (y, v) \subseteq H \times J$ and $(y, v) \star (x, u) \subseteq H \times J$. Let $(x, u) \in L$ and $(y, v) \in H \times T$ be such that $(y, v) \leq (x, u)$, ie, $y \leq_H x$, $v \leq_T u$. If $x \in I$, then $y \in I$; hence, $(y, v) \in I \times T$. Thus, $(y, v) \in L$. Similarly, if $u \in J$, then $(y, v) \in L$. Therefore, L is a hyperideal of $H \times T$. Next, we assert that $(H \times T) \setminus L$ is a subsemihypergroup of $H \times T$. Since $H \setminus I \neq \emptyset$ and $T \setminus J \neq \emptyset$, it follows that $H \setminus I$ and $T \setminus J$ are semihypergroups of H and of T, respectively. We have

$$(H \times T) \setminus L = (H \setminus I) \times (T \setminus J) \neq \emptyset.$$

Then, $(H \setminus I) \times (T \setminus J)$ is a subsemihypergroup of $H \times T$. Hence, L is a prime hyperideal of $H \times T$.

Conversely, assume that L is a prime hyperideal of $H \times T$. If $L = \emptyset$, then $L = (\emptyset \times T) \cup (H \times \emptyset)$. Assume that $(x, u) \in L$. We assert that $\{x\} \times T \subseteq L$ or $H \times \{u\} \subseteq L$. Suppose that $\{x\} \times T \nsubseteq L$ and $H \times \{u\} \nsubseteq L$. Then, there exist $v \in T$ and $y \in H$ such that $(x, v) \notin L$ and such that $(y, u) \notin L$. We have

$$(x, v) \star (y, u) \star (x, v) \star (y, u) = x \circ y \circ x \circ y \times v \diamond u \diamond v \diamond u$$

and

$$(x \circ y, v) \star (x, u) \star (y, v \diamond u) = x \circ y \circ x \circ y \times v \diamond u \diamond v \diamond u.$$

Since $(x, v) \star (y, u) \star (x, v) \star (y, u) \subseteq (H \times T) \setminus L$, it follows that

$$(x \circ y, v) \star (x, u) \star (y, v \diamond u) \subseteq (H \times T) \setminus L.$$

But, since $(x, u) \in L$, we have $(x \circ y, v) \star (x, u) \star (y, v \diamond u) \subseteq L$. This is a contradiction. Hence, $\{x\} \times T \subseteq L$ or $H \times \{u\} \subseteq L$. Let

$$A = \{x \in H \mid \{x\} \times T \subseteq L\} \quad \text{and} \quad B = \{u \in T \mid H \times \{u\} \subseteq L\},$$

and let

$$I = (A] \quad \text{and} \quad J = (B].$$

Let $(x, u) \in L$. Then, $\{x\} \times T \subseteq L$ or $H \times \{u\} \subseteq L$. Thus, $x \in I$ or $u \in J$. Hence, $(x, u) \in (I \times T) \cup (H \times J)$. Thus, $L \subseteq (I \times T) \cup (H \times J)$. The reverse inclusion is clear. Hence,

$$L = (I \times T) \cup (H \times J).$$

We shall show that I is a prime hyperideal of H. That J is a prime hyperideal of T can be proved similarly. If $I = \emptyset$, then I is a prime hyperideal of H. Assume that $I \neq \emptyset$. If $I = H$, then $L = H \times T$. This is a contradiction, since L is a prime hyperideal of $H \times T$. Hence, $H \setminus I \neq \emptyset$. Similarly, $T \setminus J \neq \emptyset$. Let $x, y \in H \setminus I$ and $u \in T \setminus J$. Then,

$$(x, u), (y, u) \in (H \setminus I) \times (T \setminus J).$$

Since L is prime, it follows that $(H \setminus I) \times (T \setminus J)$ is a subsemihypergroup of $H \times T$. Since

$$x \circ y \times u \diamond u = (x, u) \star (y, u) \subseteq (H \setminus I) \times (T \setminus J),$$

we get $x \circ y \subseteq H \setminus I$. Thus, $H \setminus I$ is a subsemihypergroup of H. Let $x \in I, y \in H$ and $u \in T \setminus J$. Since $x \in I$, it follows that $(x, u) \in L$. Since L is a hyperideal of $H \times T$, it follows that

$$(x, u) \star (y, u) = x \circ y \times u \diamond u \subseteq L$$

and

$$(y, u) \star (x, u) = y \circ x \times u \diamond u \subseteq L.$$

Since $T \setminus J$ is a subsemihypergroup, it follows that $u \diamond u \subseteq T \setminus J$. Since

$$x \circ y \times u \diamond u, y \circ x \times u \diamond u \subseteq L,$$

we obtain

$$x \circ y \times u \diamond u, y \circ x \times u \diamond u \subseteq I \times T,$$

and hence $x \circ y, y \circ x \subseteq I$. It is clear that if $x \in I$ and $y \in H$ such that $y \leq x$, then $y \in I$. Therefore, I is a prime hyperideal of H. □

Suppose that $(H, \circ)$ and $(T, \diamond)$ are semihypergroups. Then, the Cartesian product $H \times T$ is a semihypergroup under the coordinatewise multiplication. Define a partial order $\leq_H$ on H by

$$x \leq_H y \text{ if and only if } x = y \text{ for all } x, y \in H.$$

Then, H forms an ordered semihypergroup. Similarly, T forms an ordered semihypergroup with a partial order $\leq_T$ defined in a similar way. Using Theorem 3.2.2, we have the following result.

Corollary 3.2.3. *Let $(H, \circ)$ and $(T, \diamond)$ be semihypergroups. Then, a subset L of $H \times T$ is a prime hyperideal of $H \times T$ if and only if $L = (I \times T) \cup (H \times J)$ for some prime hyperideals I and J of H and of T, respectively.*

3.3 RIGHT SIMPLE ORDERED SEMIHYPERGROUPS

Let $(H, \circ, \leq)$ be an ordered semihypergroup. An element a of H is said to be *right simple* if $H = (a \circ H]$. If H contains a right simple element, then it is called a *right simple element ordered semihypergroup*. If every element of H is right simple, then H is called a *right simple ordered semihypergroup*.

Theorem 3.3.1. *If $(H, \circ, \leq_H)$ and $(T, \diamond, \leq_T)$ are two right simple element ordered semihypergroups, then $H \times T$ is a right simple element ordered semihypergroup, too. Moreover, if A and B are the sets of all right simple elements of H and of T, respectively, then $A \times B$ is the set of all right simple elements of $H \times T$.*

Proof. Assume that $(H, \circ, \leq_H)$ and $(T, \diamond, \leq_T)$ are right simple element ordered semihypergroups with the sets of all right simple elements A and B, respectively. If $(a, b) \in A \times B$, then

$$H \times T = (a \circ H] \times (b \diamond T].$$

If $(s, t) \in H \times T$, then $s \leq_H a \circ s'$ for some s' in H, and $t \leq_T b \diamond t'$ for some t' in T. Since $(s, t) \leq a \circ s' \times b \diamond t'$, it follows that

$$(s, t) \in \left(\bigcup_{(s,t) \in H \times T} a \circ s \times b \diamond t \right] = ((a, b) \star (H \times T)].$$

Hence, (a, b) is a right simple element of $H \times T$. If (a, b) is a right simple element of $H \times T$, then

$$H \times T = ((a, b) \star (H \times T)] = \left(\bigcup_{(s,t) \in H \times T} a \circ s \times b \diamond t \right].$$

If $(s, t) \in H \times T$, then $(s, t) \leq (u, v)$ for some $(u, v) \in a \circ s' \times b \diamond t'$ where $s' \in H, t' \in T$. Since

$$s \leq u \in a \circ s' \subseteq (a \circ H],$$

we have $s \in (a \circ H]$, and so $H = (a \circ H]$. Similarly, $T = (b \diamond T]$. Hence, $(a, b) \in A \times B$. □

The following result is well known in semigroup theory. Let H be a right simple element semigroup and let R denote the set of all right simple elements of H. Then, the following conditions hold: (1) R is a subsemigroup of H; (2) If $H \setminus R$ is non-empty, then it is the maximal right ideal of H and is prime, too [99]. In the following, we extend the above result based on ordered semihypergroups.

Theorem 3.3.2. *Let $(H, \circ, \leq)$ be right simple element ordered semihypergroup with the set of all right simple elements R. The following statements hold:*
(1) *R is a subsemihypergroup of H.*
(2) *If $H \setminus R$ is non-empty, then it is the maximal right hyperideal of H and is prime, too.*

Proof. If $(H, \circ, \leq)$ is a right simple ordered semihypergroup, then it is clear that (1) and (2) hold. Then, we assume that $(H, \circ, \leq)$ is not a right simple ordered semihypergroup.
(1) Let $a, b \in R$. Since $H = (a \circ H]$ and $H = (b \circ H]$, it follows that

$$H = (a \circ H] = (a \circ (b \circ H]] \subseteq ((a] \circ (b \circ H]] = (a \circ b \circ H],$$

and so $a \circ b \subseteq R$.
(2) Assume that $H \setminus R \neq \emptyset$. Let $x \in H$ and $a \in H \setminus R$. If $a \circ x \subseteq R$, then $H = (a \circ x \circ H] \subseteq (a \circ H]$; hence, $a \in R$. This is a contradiction. Thus, $a \circ x \subseteq H \setminus R$. Let $x \in H \setminus R$ and $y \in H$ be such that $y \leq x$. If $y \in R$, then $H = (y \circ H] \subseteq (x \circ H]$; hence, $x \in R$. This is a contradiction. Thus, $H \setminus R$ is a right hyperideal of H. Let A be a right hyperideal of H such that $H \setminus R \subset A$. Then, there is an element a in $A \setminus (H \setminus R)$. Since $H = (a \circ H] \subseteq A$, it follows that $A = H$. By (1), it follows directly that $H \setminus R$ is prime.

□

Theorem 3.3.3. *If an ordered semihypergroup $(H, \circ, \leq)$ has a unique maximal right hyperideal A such that $H \setminus A \neq (b]$ for all b in $H \setminus A$, then $H \setminus A$ is the set of all right simple elements of H.*

Proof. Let R denote the set of all right simple elements of H. Let $a \in R$. If $a \in A$, then $H = (a \circ H] \subseteq A$. This is a contradiction. Thus, $R \subseteq H \setminus A$. Let $b \in H \setminus A$. We have, $(b \circ H]$ is a right hyperideal of H. If $(b \circ H] \subset H$, then by assumption we have $(b \circ H] \subseteq A$; hence, $(A \cup \{b\}]$ is a right hyperideal of H. By $A \subset (A \cup \{b\}]$, $H = (A \cup \{b\}]$. Hence, $H \setminus A = (b]$. This is a contradiction. Hence, $(b \circ H] = H$. □

Let $(H, \circ, \leq)$ be an ordered semihypergroup. An equivalence relation $\mathcal{R}$ is defined on H by

$$a\mathcal{R}b \text{ if and only if } (a \cup a \circ H] = (b \cup b \circ H]$$

for any a, b in H.

An element a of an ordered semihypergroup $(H, \circ, \leq)$ is said to be *right regular* if $a \in (a^2 \circ H]$.

Theorem 3.3.4. *Let $(H, \circ, \leq)$ be a right simple element ordered semihypergroup with a set of all right simple elements R. Then*

(1) *R is an $\mathcal{R}$-class of H;*
(2) *every element of R is right regular.*

Proof

(1) If $a, b \in R$, then $H = (a \circ H]$ and $H = (b \circ H]$; hence, $(a \cup a \circ H] = (b \cup b \circ H]$. This shows that $a\mathcal{R}b$. Let $x \in H$ be such that $x\mathcal{R}a$ for some a in R. Then,

$$(x \cup x \circ H] = (a \cup a \circ H] = H.$$

If $H \setminus R = \emptyset$, then $x \in R$. If $H \setminus R \neq \emptyset$ and $x \in H \setminus R$, then

$$H = (x \cup x \circ H] \subseteq H \setminus R;$$

hence $H = H \setminus R$. This is a contradiction. Hence, $x \in R$.

(2) If $a \in R$, then

$$a \in (a \circ H] \subseteq (a \circ (a \circ H]] \subseteq (a^2 \circ H];$$

hence a is right regular.

□

3.4 ORDERED SEMIGROUPS (SEMIHYPERGROUPS) DERIVED FROM ORDERED SEMIHYPERGROUPS

Now, the following question is natural: If $(H, \circ, \leq)$ is an ordered semihypergroup and ρ is a strongly regular relation on H, then is the set H/ρ an ordered semigroup? Similar to ordered semigroups, a probable order on H/ρ could be the relation $\preceq$ on H/ρ defined by means of the order $\leq$ on H, that is,

$$\preceq := \{(\rho(a), \rho(b)) \in H/\rho \times H/\rho \mid \exists x \in \rho(a), \exists y \in \rho(b) \text{ such that } (x, y) \in \leq\}.$$

But this relation is not an order in general. The following question arises:

Question. Is there a strongly regular relation ρ on H for which H/ρ is an ordered semigroup?

Our main aim in this section is to answer the above question. The main reference is [100].

We begin with the following definition.

Definition 3.4.1. Let $(H, \circ, \leq)$ be an ordered semihypergroup. A relation ρ on H is called *pseudoorder* if

(1) $\leq \subseteq \rho$;
(2) $a\rho b$ and $b\rho c$ imply $a\rho c$;
(3) $a\rho b$ implies $a \circ c \overline{\overline{\rho}} b \circ c$ and $c \circ a \overline{\overline{\rho}} c \circ b$, for all $c \in H$.

Theorem 3.4.2. *Let $(H, \circ, \leq)$ be an ordered semihypergroup and ρ be a pseudoorder on H. Then, there exists a strongly regular relation ρ^* on H such that H/ρ^* is an ordered semigroup.*

Proof. Suppose that ρ^* is the relation on H defined as follows:

$$\rho^* = \{(a, b) \in H \times H \mid a\rho b \text{ and } b\rho a\}.$$

First, we show that ρ^* is a strongly regular relation on H. Let a be an arbitrary element of H. Clearly, $(a, a) \in \leq \subseteq \rho$, so $a\rho^* a$. If $(a, b) \in \rho^*$, then $a\rho b$, and $b\rho a$. Hence, $(b, a) \in \rho^*$. If $(a, b) \in \rho^*$ and $(b, c) \in \rho^*$, then $a\rho b$, $b\rho a$, $b\rho c$ and $c\rho b$. Hence, $a\rho c$ and $c\rho a$, which imply that $(a, c) \in \rho^*$. Thus, ρ^* is an equivalence relation. Now, let $a\rho^* b$ and $c \in H$. Then, $a\rho b$ and $b\rho a$. Since ρ is pseudoorder on H, by condition (3) of Definition 3.4.1, we conclude that

$$a \circ c \overline{\overline{\rho}} b \circ c, c \circ a \overline{\overline{\rho}} c \circ b,$$

$$b \circ c \overline{\overline{\rho}} a \circ c, c \circ b \overline{\overline{\rho}} c \circ a.$$

Hence, for every $x \in a \circ c$ and $y \in b \circ c$, we have $x\rho y$ and $y\rho x$ which imply that $x\rho^* y$. So, $a \circ c \overline{\overline{\rho^*}} b \circ c$. Similarly, we obtain $c \circ a \overline{\overline{\rho^*}} c \circ b$. Thus, ρ^* is a strongly regular relation on H. Hence, by Theorem 1.5.3, H/ρ^* with the following operation is a semigroup:

$$\rho^*(x) \odot \rho^*(y) = \rho^*(z), \text{ for all } z \in x \circ y.$$

Now, we define a relation $\preceq$ on H/ρ^* as follows:

$$\preceq := \{(\rho^*(x), \rho^*(y)) \in H/\rho^* \times H/\rho^* \mid \exists a \in \rho^*(x), \exists b \in \rho^*(y) \text{ such that } (a, b) \in \rho\}.$$

We show that

$$\rho^*(x) \preceq \rho^*(y) \Leftrightarrow x\rho y.$$

Let $\rho^*(x) \preceq \rho^*(y)$. We show that for every $a \in \rho^*(x)$ and $b \in \rho^*(y)$, $a\rho b$. Since $\rho^*(x) \preceq \rho^*(y)$, there exist $x' \in \rho^*(x)$ and $y' \in \rho^*(y)$ such that $x'\rho y'$. Since $a \in \rho^*(x)$ and $x' \in \rho^*(x)$, we obtain $a\rho^* x'$, and so $a\rho x'$ and $x'\rho a$. Since $b \in \rho^*(y)$ and $y' \in \rho^*(y)$, we obtain $b\rho^* y'$, and so $b\rho y'$ and $y'\rho b$. Now, we have $a\rho x'$, $x'\rho y'$ and $y'\rho b$, which imply that $a\rho b$. Since $x \in \rho^*(x)$ and $y \in \rho^*(y)$, we conclude that $x\rho y$. Conversely, let $x\rho y$. Since $x \in \rho^*(x)$ and $y \in \rho^*(y)$, clearly, we obtain $\rho^*(x) \preceq \rho^*(y)$.

Finally, we prove that $H/\rho^*, \odot, \preceq)$ is an ordered semigroup. Suppose that $\rho^*(x) \in H/\rho^*$, where $x \in H$. Then, $(x, x) \in \leq \subseteq \rho$. Hence, $\rho^*(x) \preceq$

$\rho^*(x)$. Let $\rho^*(x) \preceq \rho^*(y)$ and $\rho^*(y) \preceq \rho^*(x)$. Then, $x\rho y$ and $y\rho x$. Thus, $x\rho^* y$, which means that $\rho^*(x) = \rho^*(y)$. Now, let $\rho^*(x) \preceq \rho^*(y)$ and $\rho^*(y) \preceq \rho^*(z)$. Then, $x\rho y$ and $y\rho z$. So, $x\rho z$. This implies that $\rho^*(x) \preceq \rho^*(z)$.

Now, let $\rho^*(x) \preceq \rho^*(y)$ and $\rho^*(z) \in H/\rho^*$. Then, $x\rho y$ and $z \in H$. By condition (3) of Definition 3.4.1, we have $x \circ z \overline{\overline{\rho}} y \circ z$ and $z \circ x \overline{\overline{\rho}} z \circ y$. So, for all $a \in x \circ z$ and $b \in y \circ z$, we have $a\rho b$. This implies that $\rho^*(a) \preceq \rho^*(b)$. Hence, $\rho^*(x) \odot \rho^*(z) \preceq \rho^*(y) \odot \rho^*(z)$. Similarly, we get $\rho^*(z) \odot \rho^*(x) \preceq \rho^*(z) \odot \rho^*(y)$. □

Example 58. Suppose that $H = \{a, b, c, d, e\}$. We consider the ordered semihypergroup $(H, \circ, \leq)$, where the hyperoperation $\circ$ is defined by the following table:

$\circ$	a	b	c	d	e
a	a	$\{a, b\}$	$\{a, c\}$	$\{a, d\}$	e
b	a	$\{a, b\}$	$\{a, c\}$	$\{a, d\}$	e
c	a	$\{a, b\}$	$\{a, c\}$	$\{a, d\}$	e
d	a	$\{a, b\}$	$\{a, c\}$	$\{a, d\}$	e
e	a	$\{a, b\}$	$\{a, c\}$	$\{a, d\}$	e

and the order $\leq$ is defined by:

$$\leq := \{(a, a), (b, b), (c, c), (d, d), (e, e), (b, a), (c, a), (d, a)\}.$$

The covering relation and the figure of H are given by

$$\prec = \{(b, a), (c, a), (d, a)\}.$$

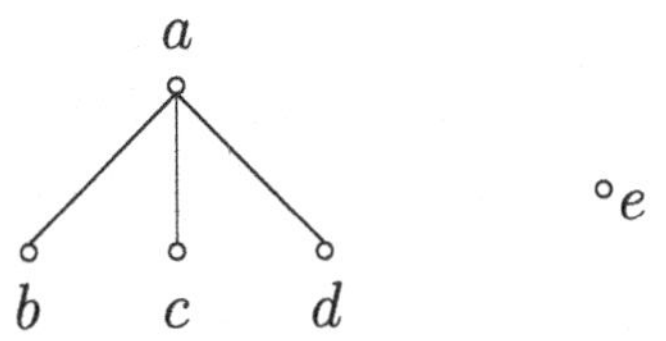

Let ρ be pseudoorder on H, defined as follows:

$$\rho = \{(a, a), (b, b), (c, c), (d, d), (e, e), (a, b), (b, a), (a, c), (c, a), (a, d), (d, a), (b, c), (c, b), (b, d), (d, b), (c, d), (d, c), (e, a), (e, b), (e, c), (e, d)\}.$$

Then, by the definition of ρ^*, we get

$$\rho^* = \{(a,a),(b,b),(c,c),(d,d),(e,e),(a,b),(b,a),\\ (a,c),(c,a),(a,d),(d,a),(b,c),(c,b),(b,d),\\ (d,b),(c,d),(d,c)\}.$$

Hence, $H/\rho^* = \{u_1, u_2\}$, where $u_1 = \{a,b,c,d\}$ and $u_2 = \{e\}$. Now, $(H/\rho^*, \odot, \preceq)$ is an ordered semigroup, where $\odot$ is defined in the following table:

$\odot$	u_1	u_2
u_1	u_1	u_2
u_2	u_1	u_2

and $\preceq = \{(u_1,u_1),(u_2,u_1),(u_2,u_2)\}$.

Theorem 3.4.3. *Let $(H, \circ, \leq)$ be an ordered semihypergroup and ρ be a pseudoorder on H. Let*

$$\mathcal{X} = \{\theta | \theta \text{ be pseudoorder on } H \text{ such that } \rho \subseteq \theta\}.$$

Let $\mathcal{Y}$ be the set of all pseudoorders on H/ρ^*. Then, card($\mathcal{X}$)=card($\mathcal{Y}$).

Proof. For $\theta \in \mathcal{X}$, we define a relation θ' on H/ρ^* as follows:

$$\theta' := \{(\rho^*(x), \rho^*(y)) \in H/\rho^* \times H/\rho^* | \exists a \in \rho^*(x), \exists b \in \rho^*(y) \text{ such that } (a,b) \in \theta\}.$$

First, we show that

$$(\rho^*(x), \rho^*(y)) \in \theta' \Leftrightarrow (x,y) \in \theta.$$

Let $(\rho^*(x), \rho^*(y)) \in \theta'$. We show that for every $a \in \rho^*(x)$ and $b \in \rho^*(y)$, $(a,b) \in \theta$. Since $(\rho^*(x), \rho^*(y)) \in \theta'$, there exist $x' \in \rho^*(x)$ and $y' \in \rho^*(y)$ such that $(x', y') \in \theta$. Since $a\rho^*x$ and $x\rho^*x'$, we have $a\rho^*x'$. So, $a\rho x'$. Since $\rho \subseteq \theta$, it follows that $a\theta x'$. Similarly, we obtain $y'\theta b$. Now, we have $a\theta x'$, $x'\theta y'$, and $y'\theta b$. Thus, $a\theta b$. Since $x \in \rho^*(x)$ and $y \in \rho^*(y)$, we conclude that $(x,y) \in \theta$. Conversely, let $(x,y) \in \theta$. Since $x \in \rho^*(x)$ and $y \in \rho^*(y)$, clearly, we obtain $(\rho^*(x), \rho^*(y)) \in \theta'$.

Now, let $(\rho^*(x), \rho^*(y)) \in \preceq$. Then, by Theorem 3.4.2, $(x,y) \in \rho \subseteq \theta$. This implies that $(\rho^*(x), \rho^*(y)) \in \theta'$. Hence, $\preceq \subseteq \theta'$. Now, suppose that $(\rho^*(x), \rho^*(y)) \in \theta'$ and $(\rho^*(y), \rho^*(z)) \in \theta'$. Then, $(x,y) \in \theta$ and $(y,z) \in \theta$, which imply that $(x,z) \in \theta$. Thus, $(\rho^*(x), \rho^*(z)) \in \theta'$. Also, if $(\rho^*(x), \rho^*(y)) \in \theta'$ and $\rho^*(z) \in H/\rho^*$, then $(x,y) \in \theta$ and $z \in H$. Then, $x \circ z \overline{\overline{\theta}} y \circ z$ and $z \circ x \overline{\overline{\theta}} z \circ y$. So, for all $a \in x \circ z$ and for all $b \in y \circ z$, we have $a\theta b$. This implies that $\theta'(\rho^*(a)) = \theta'(\rho^*(b))$ and so $\theta'(\rho^*(x) \odot$

$\rho^*(z)) = \theta'(\rho^*(y) \odot \rho^*(z))$. Thus, $(\rho^*(x) \odot \rho^*(z))\theta'(\rho^*(y) \odot \rho^*(z))$. Similarly, we obtain $(\rho^*(z) \odot \rho^*(x))\theta'(\rho^*(z) \odot \rho^*(y))$. Therefore, if $\theta \in \mathcal{X}$, then θ' is a pseudoorder on H/ρ^*.

Now, we define the map $\psi : \mathcal{X} \to \mathcal{Y}$ by $\psi(\theta) = \theta'$.

Let $\theta_1, \theta_2 \in \mathcal{X}$ and $\theta_1 = \theta_2$. Suppose that $(\rho^*(x), \rho^*(y)) \in \theta'_1$ is an arbitrary element. Then, $(x, y) \in \theta_1$ and so $(x, y) \in \theta_2$. This implies that $(\rho^*(x), \rho^*(y)) \in \theta'_2$. Thus, $\theta'_1 \subseteq \theta'_2$. Similarly, we obtain $\theta'_2 \subseteq \theta'_1$. Therefore, ψ is well defined.

Let $\theta_1, \theta_2 \in \mathcal{X}$ and $\theta'_1 = \theta'_2$. Suppose that $(x, y) \in \theta_1$ is an arbitrary element. Then, $(\rho^*(x), \rho^*(y)) \in \theta'_1$, and so $(\rho^*(x), \rho^*(y)) \in \theta'_2$. This implies that $(x, y) \in \theta_2$. Thus, $\theta_1 \subseteq \theta_2$. Similarly, we obtain $\theta_2 \subseteq \theta_1$. Therefore, ψ is one-to-one.

Finally, we prove that ψ is onto. Consider $\Sigma \in \mathcal{Y}$. We define a relation θ on H as follows:

$$\theta = \{(x, y) | (\rho^*(x), \rho^*(y)) \in \Sigma\}.$$

We show that θ is a pseudoorder on H and $\rho \subseteq \theta$. Suppose that $(x, y) \in \rho$. By Theorem 3.4.2, $(\rho^*(x), \rho^*(y)) \in \preceq \subseteq \Sigma$, and so $(x, y) \in \theta$. If $(x, y) \in \leq$, then $(x, y) \in \rho \subseteq \theta$. Hence, $\leq \subseteq \theta$. Let $(x, y) \in \theta$ and $(y, z) \in \theta$. Then, $(\rho^*(x), \rho^*(y)) \in \Sigma$ and $(\rho^*(y), \rho^*(z)) \in \Sigma$. So, $(\rho^*(x), \rho^*(z)) \in \Sigma$. This implies that $(x, z) \in \theta$.

Now, let $(x, y) \in \theta$ and $z \in H$. Then, $(\rho^*(x), \rho^*(y)) \in \Sigma$ and $\rho^*(z) \in H/\rho^*$. Thus, $(\rho^*(x) \odot \rho^*(z), \rho^*(y) \odot \rho^*(z)) \in \Sigma$. Therefore, for all $a \in x \circ z$ and for all $b \in y \circ z$, we have $(\rho^*(a), \rho^*(b)) \in \Sigma$ and this means that $(a, b) \in \theta$. Therefore, $x \circ z \overline{\overline{\theta}} y \circ z$. Similarly, we obtain $z \circ x \overline{\overline{\theta}} z \circ y$. Now, obviously we have $\theta' = \Sigma$. □

Remark 15. In Theorem 3.4.3, it is easy to see that $\theta_1 \subseteq \theta_2$ if and only if $\theta'_1 \subseteq \theta'_2$.

Remark 16. If $(H, \circ, \leq_H)$ and $(T, \diamond, \leq_T)$ are two ordered semigroups and $\varphi : H \to T$ is a homomorphism, we denote by k, the pseudoorder on H defined by $k = \{(a, b) | \varphi(a) \leq_T \varphi(b)\}$. Then, we have $ker\varphi = k^*$.

Corollary 3.4.4. *Let $(H, \circ, \leq_H)$ and $(T, \diamond, \leq_T)$ be two ordered semigroups and $\varphi : H \to T$ be a homomorphism. Then, $H/ker\varphi \cong Im\varphi$.*

Let $(H, \circ, \leq_H)$ be an ordered semihypergroup, and let ρ, θ be pseudoorders on H such that $\rho \subseteq \theta$. We define a relation θ/ρ on H/ρ^* as follows:

$$\theta/\rho := \{(\rho^*(a), \rho^*(b)) \in H/\rho^* \times H/\rho^* | \exists x \in \rho^*(a), \exists y \in \rho^*(b) \text{ such that } (x, y) \in \theta\}.$$

Then, we can see that

$$(\rho^*(a), \rho^*(b)) \in \theta/\rho \Leftrightarrow (a, b) \in \theta.$$

Theorem 3.4.5. *Let $(H, \circ, \preceq_H)$ be an ordered semihypergroup, and let ρ, θ be pseudoorders on H such that $\rho \subseteq \theta$. Then,*

(1) *θ/ρ is a pseudoorder on H/ρ^*;*

(2) *$(H/\rho^*)/(\theta/\rho)^* \cong H/\theta^*$.*

Proof

(1) If $(\rho^*(a), \rho^*(b)) \in \preceq_\rho$, then $(a, b) \in \rho$. So, $(a, b) \in \theta$ which implies that $(\rho^*(a), \rho^*(b)) \in \theta/\rho$. Thus, $\preceq_\rho \subseteq \theta/\rho$. Let $(\rho^*(a), \rho^*(b)) \in \theta/\rho$ and $(\rho^*(b), \rho^*(c)) \in \theta/\rho$. Then, $(a, b) \in \theta$ and $(b, c) \in \theta$. Hence, $(a, c) \in \theta$ and so $(\rho^*(a), \rho^*(c)) \in \theta/\rho$. Now, let $(\rho^*(a), \rho^*(b)) \in \theta/\rho$ and $\rho^*(c) \in H/\rho^*$. Then, $(a, b) \in \theta$. Since θ is pseudoorder on H, we obtain $a \circ c \overline{\overline{\theta}} b \circ c$ and $c \circ a \overline{\overline{\theta}} c \circ b$. Hence, for all $x \in a \circ c$ and for all $y \in b \circ c$, we have $(x, y) \in \theta$. This implies that $(\rho^*(x), \rho^*(y)) \in \theta/\rho$. Since ρ^* is a strongly regular relation on H, $\rho^*(x) = \rho^*(a) \odot \rho^*(c)$ and $\rho^*(y) = \rho^*(b) \odot \rho^*(c)$. So, we obtain $(\rho^*(a) \odot \rho^*(c), \rho^*(b) \odot \rho^*(c)) \in \theta/\rho$. Similarly, we obtain $(\rho^*(c) \odot \rho^*(a), \rho^*(c) \odot \rho^*(b)) \in \theta/\rho$. Therefore, θ/ρ is a pseudoorder on H/ρ^*.

(2) We define the map $\psi : H/\rho^* \to H/\theta^*$ by $\psi(\rho^*(a)) = \theta^*(a)$. If $\rho^*(a) = \rho^*(b)$, then $(a, b) \in \rho^*$. Hence, by the definition of ρ^*, $(a, b) \in \rho \subseteq \theta$ and $(b, a) \in \rho \subseteq \theta$. This implies that $(a, b) \in \theta^*$ and so $\theta^*(a) = \theta^*(b)$. Thus, θ is well-defined. For all $\rho^*(x), \rho^*(y) \in H/\rho^*$, we have

$$\rho^*(x) \odot \rho^*(y) = \rho^*(z), \text{ for all } z \in x \circ y,$$
$$\theta^*(x) \otimes \theta^*(y) = \theta^*(z), \text{ for all } z \in x \circ y.$$

Thus,

$$\begin{aligned}\psi(\rho^*(x) \odot \rho^*(y)) &= \psi(\rho^*(z)), \text{ for all } z \in x \circ y \\ &= \theta^*(z), \text{ for all } z \in x \circ y \\ &= \theta^*(x) \otimes \theta^*(y) \\ &= \psi(\rho^*(x)) \otimes \psi(\rho^*(y)),\end{aligned}$$

and if $\rho^*(x) \preceq_\rho \rho^*(y)$, then $(x, y) \in \rho$. So, $(x, y) \in \theta$ and this implies that $\theta^*(x) \preceq_\theta \theta^*(y)$. Therefore, ψ is a homomorphism. It is easy to see that ψ is onto mapping, since

$$Im\psi = \{\psi(\rho^*(x)) | x \in H\} = \{\theta^*(x) | x \in H\} = H/\theta^*.$$

So, by Corollary 3.4.4, we obtain

$$(H/\rho^*)/ker\psi \cong Im\psi = H/\theta^*.$$

Suppose that

$$k := \{(\rho^*(x), \rho^*(y)) | \psi(\rho^*(x)) \preceq_\theta \psi(\rho^*(y))\}.$$

Then,

$$\begin{aligned}(\rho^*(x), \rho^*(y)) \in k &\Leftrightarrow \psi(\rho^*(x)) \preceq_\theta \psi(\rho^*(y)) \\ &\Leftrightarrow \theta^*(x) \preceq_\theta \theta^*(y) \\ &\Leftrightarrow (x, y) \in \theta \\ &\Leftrightarrow (\rho^*(x), \rho^*(y)) \in \theta/\rho.\end{aligned}$$

Hence, $k = \theta/\rho$ and by Remark 16, we have $k^* = (\theta/\rho)^* = ker\psi$.

□

Definition 3.4.6. Let $(H, \circ, \leq_H)$ and $(T, \diamond, \leq_T)$ be two ordered semihypergroups, ρ_1, ρ_2 be two pseudoorders on H, T, respectively, and the map $f : H \to T$ be a homomorphism. Then, f is called a (ρ_1, ρ_2)-*homomorphism* if

$$(x, y) \in \rho_1 \Rightarrow (f(x), f(y)) \in \rho_2.$$

Lemma 3.4.7. *Let $(H, \circ, \leq_H)$ and $(T, \diamond, \leq_T)$ be two ordered semihypergroups, ρ_1, ρ_2 be two pseudoorders on H, T, respectively, and the map $f : H \to T$ be a (ρ_1, ρ_2)-homomorphism. Then, the map $\overline{f} : H/\rho_1^* \to T/\rho_2^*$ defined by*

$$\overline{f}(\rho_1^*(x)) = \rho_2^*(f(x)), \textit{ for all } x \in H$$

is a homomorphism of semigroups.

Proof. Suppose that $\rho_1^*(x) = \rho_1^*(y)$. Then, $(x, y) \in \rho_1$ and $(y, x) \in \rho_1$. Since f is a (ρ_1, ρ_2)-homomorphism, it follows that $(f(x), f(y)) \in \rho_2$ and $(f(y), f(x)) \in \rho_2$. This implies that $\rho_2^*(f(x)) = \rho_2^*(f(y))$ or $\overline{f}(\rho_1^*(x)) = \overline{f}(\rho_1^*(y))$. Therefore, $\overline{f}$ is well defined. Now, we show that $\overline{f}$ is a homomorphism. Suppose that $\rho_1^*(x)$, $\rho_1^*(y)$ be two arbitrary elements of H/ρ_1^*. Then,

$$\overline{f}(\rho_1^*(x) \odot \rho_1^*(y)) = \overline{f}(\rho_1^*(z)) = \rho_2^*(f(z)), \text{ for all } z \in x \circ y.$$

Since $z \in x \circ y$, it follows that $f(z) \in f(x) \diamond f(y)$. Since ρ_2^* is a strongly regular relation, we obtain $\rho_2^*(f(z)) = \rho_2^*(f(x)) \otimes \rho_2^*(f(y))$. Thus, we have

$$\overline{f}(\rho_1^*(x) \odot \rho_1^*(y)) = \rho_2^*(f(x)) \otimes \rho_2^*(f(y)) = \overline{f}(\rho_1^*(x)) \otimes \overline{f}(\rho_1^*(y)).$$

□

Theorem 3.4.8. *Let $(H, \circ, \preceq_H)$ and $(T, \diamond, \preceq_T)$ be two ordered semihypergroups, ρ_1, ρ_2 be two pseudoorders on H, T, respectively, and the map $f : H \to T$ be a (ρ_1, ρ_2)-homomorphism. Then, the relation ρ_f defined by*

$$(\rho_1)_f := \{(\rho_1^*(x), \rho_1^*(y)) | \rho_2^*(f(x)) \preceq_T \rho_2^*(f(y))\}$$

is a pseudoorder on H/ρ_1^*.

Proof. Suppose that $(\rho_1^*(x), \rho_1^*(y)) \in \preceq_H$. Since $\rho_1^*(x) \preceq_H \rho_1^*(y)$ and $\overline{f}$ is a homomorphism (by Lemma 3.4.7), we have $\overline{f}(\rho_1^*(x)) \preceq_T \overline{f}(\rho_1^*(y))$. Thus, $\rho_2^*(f(x)) \preceq_T \rho_2^*(f(y))$. This means that $(\rho_1^*(x), \rho_1^*(y)) \in (\rho_1)_f$.

Let $(\rho_1^*(x), \rho_1^*(y)) \in (\rho_1)_f$ and $(\rho_1^*(y), \rho_1^*(z)) \in (\rho_1)_f$. Then, we have $\rho_2^*(f(x)) \preceq_T \rho_2^*(f(y))$ and $\rho_2^*(f(y)) \preceq_T \rho_2^*(f(z))$. So, $\rho_2^*(f(x)) \preceq_T \rho_2^*(f(z))$. This implies that $(\rho_1^*(x), \rho_1^*(z)) \in (\rho_1)_f$. Now, let $(\rho_1^*(x), \rho_1^*(y)) \in (\rho_1)_f$ and $\rho_1^*(z) \in H/\rho_1^*$. We show that $(\rho_1^*(x) \odot \rho_1^*(z), \rho_1^*(y) \odot \rho_1^*(z)) \in (\rho_1)_f$. Since $(\rho_1^*(x), \rho_1^*(y)) \in (\rho_1)_f$, it follows that $\overline{f}(\rho_1^*(x)) \preceq_T \overline{f}(\rho_1^*(y))$. Thus, by definition, we get $\overline{f}(\rho_1^*(x)) \otimes \overline{f}(\rho_1^*(z)) \preceq_T \overline{f}(\rho_1^*(y)) \otimes \overline{f}(\rho_1^*(z))$. So, $\overline{f}(\rho_1^*(x) \odot \rho_1^*(z)) \preceq_T \overline{f}(\rho_1^*(y) \odot \rho_1^*(z))$. Hence, for all $u \in x \circ z$ and for all $v \in y \circ z$, we have $\overline{f}(\rho_1^*(u)) \preceq_T \overline{f}(\rho_1^*(v))$. This implies that $(\rho_1^*(u), \rho_1^*(v)) \in (\rho_1)_f$. Therefore, we have $(\rho_1^*(x) \odot \rho_1^*(z), \rho_1^*(y) \odot \rho_1^*(z)) \in (\rho_1)_f$. Similarly, we obtain $(\rho_1^*(z) \odot \rho_1^*(x), \rho_1^*(z) \odot \rho_1^*(y)) \in (\rho_1)_f$. □

Corollary 3.4.9. *$ker\overline{f} = (\rho_1)_f^*$.*

Proof. It is straightforward. □

Corollary 3.4.10. *Let $(H, \circ, \preceq_H)$ and $(T, \diamond, \preceq_T)$ be two ordered semihypergroups, ρ_1, ρ_2 be two pseudoorders on H, T, respectively, and the map $f : H \to T$ be a (ρ_1, ρ_2)-homomorphism. Then, the following diagram is commutative.*

$$\begin{array}{ccc} H & \xrightarrow{\quad f \quad} & T \\ {\scriptstyle \varphi_H}\big\downarrow & & \big\downarrow{\scriptstyle \varphi_T} \\ H/\rho_1^* & \xrightarrow[\quad \overline{f} \quad]{} & T/\rho_2^* \end{array}$$

Proof. It is straightforward. □

Theorem 3.4.11. *Let $(H, \circ, \preceq_H)$ and $(T, \diamond, \preceq_T)$ be two ordered semihypergroups, ρ_1, ρ_2 be two pseudoorders on H, T, respectively, and the map $f : H \to T$ be a (ρ_1, ρ_2)-homomorphism. If Σ is a pseudoorder on H/ρ_1^* such that $\Sigma \subseteq (\rho_1)_f$, then the mapping $\psi : (H/\rho_1^*)/\Sigma^* \to T/\rho_2^*$ defined by $\psi(\Sigma^*(\rho_1^*(x))) = \overline{f}(\rho_1^*(x))$ is the unique homomorphism of $(H/\rho_1^*)/\Sigma^*$ into T/ρ_2^* such that the following diagram is commutative.*

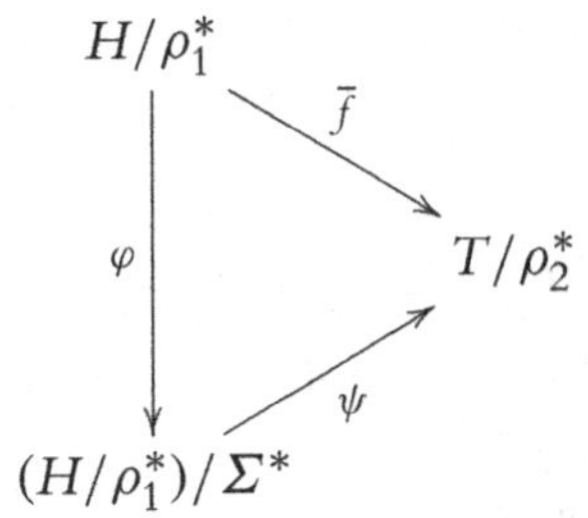

Conversely, if Σ is a pseudoorder on H/ρ_1^* for which there exists a homomorphism $\psi : (H/\rho_1^*)/\Sigma^* \to T/\rho_2^*$ such that the above diagram commutes, then $\Sigma \subseteq (\rho_1)_f$.

Proof. Suppose that Σ is a pseudoorder on H/ρ_1^* and $\Sigma \subseteq (\rho_1)_f$. First, we show that ψ is well-defined. If $\Sigma^*(\rho_1^*(x)) = \Sigma^*(\rho_1^*(y))$, then $(\rho_1^*(x), \rho_1^*(y)) \in \Sigma^*$. So, $(\rho_1^*(x), \rho_1^*(y)) \in \Sigma \subseteq (\rho_1)_f$ and $(\rho_1^*(y), \rho_1^*(x)) \in \Sigma \subseteq (\rho_1)_f$. Thus, $\rho_2^*(f(x)) \preceq_T \rho_2^*(f(y))$ and $\rho_2^*(f(y)) \preceq_T \rho_2^*(f(x))$, and so $\rho_2^*(f(x)) = \rho_2^*(f(y))$. Therefore, $\overline{f}(\rho_1^*(x)) = \overline{f}(\rho_1^*(y))$. Now, let $\rho_1^*(x)$ and $\rho_1^*(y)$ be two arbitrary elements of H/ρ_1^*. Then,

$$\begin{aligned}\psi(\Sigma^*(\rho_1^*(x))\bigstar\Sigma^*(\rho_1^*(y))) &= \psi(\Sigma^*(\rho_1^*(x) \odot \rho_1^*(y))\\ &= \overline{f}(\rho_1^*(x) \odot \rho_1^*(y))\\ &= \overline{f}(\rho_1^*(x)) \otimes \overline{f}(\rho_1^*(y))\\ &= \psi(\Sigma^*(\rho_1^*(x))) \otimes \psi(\Sigma^*(\rho_1^*(y))),\end{aligned}$$

and if $\Sigma^*(\rho_1^*(x)) \preceq_\Sigma \Sigma^*(\rho_1^*(y))$, then $(\rho_1^*(x), \rho_1^*(y)) \in \Sigma \subseteq (\rho_1)_f$. This implies that $\overline{f}(\rho_1^*(x)) \preceq_T \overline{f}(\rho_1^*(y)))$, and hence, $\psi(\Sigma^*(\rho_1^*(x))) \preceq_T \psi(\Sigma^*(\rho_1^*(y)))$. Therefore, ψ is a homomorphism. Clearly, we have $\psi\varphi = \overline{f}$.

For the converse of theorem, we have

$$\begin{aligned}(\rho_1^*(x), \rho_1^*(y)) \in \Sigma &\Rightarrow \Sigma^*(\rho_1^*(x)) \preceq_\Sigma \Sigma^*(\rho_1^*(y))\\ &\Rightarrow \psi(\Sigma^*(\rho_1^*(x))) \preceq_T \psi(\Sigma^*(\rho_1^*(y)))\\ &\Rightarrow \psi(\varphi(\rho_1^*(x))) \preceq_T \psi(\varphi(\rho_1^*(y)))\\ &\Rightarrow \overline{f}(\rho_1^*(x)) \preceq_T \overline{f}(\rho_1^*(y))\\ &\Rightarrow \rho_2^*(f(x)) \preceq_T \rho_2^*(f(y))\\ &\Rightarrow (\rho_1^*(x), \rho_1^*(y)) \in (\rho_1)_f.\end{aligned}$$

□

Definition 3.4.12. Let $(H, \circ, \leq_H)$ and $(T, \diamond, \leq_T)$ be two ordered semihypergroups, and let ρ_1, ρ_2 be two pseudoorders on H, T, respectively. On $H \times T$ we define

$$(s_1, t_1)\rho(s_2, t_2) \Leftrightarrow s_1\rho_1 s_2 \text{ and } t_1\rho_2 t_2.$$

Lemma 3.4.13. *In Definition 3.4.12, ρ is pseudoorder on $H \times T$.*

Proof. It is straightforward. □

Theorem 3.4.14. *Let $(H, \circ, \leq_H)$ and $(T, \diamond, \leq_T)$ be two ordered semihypergroups, ρ_1, ρ_2 be two pseudoorders on H, T, respectively. Then,*

$$(H \times T)/\rho^* \cong H/\rho_1^* \times T/\rho_2^*.$$

Proof. We consider the map $\psi : (H \times T)/\rho^* \to H/\rho_1^* \times T/\rho_2^*$ by $\psi(\rho^*(s, t)) = (\rho_1^*(s), \rho_2^*(t))$. Suppose that $\rho^*(s_1, t_1) = \rho^*(s_2, t_2)$. Then, $(s_1, t_1)\rho^*(s_2, t_2)$ which implies that $(s_1, t_1)\rho(s_2, t_2)$ and $(s_2, t_2)\rho(s_1, t_1)$. Hence, $s_1\rho_1 s_2$, $t_1\rho_2 t_2$, $s_2\rho_1 s_1$, and $t_2\rho_2 t_1$, which imply that $s_1\rho_1^* s_2$ and $t_1\rho_2^* t_2$. So, $(\rho_1^*(s_1), \rho_2^*(t_1)) = (\rho_1^*(s_2), \rho_2^*(t_2))$. This means that $\psi(\rho^*(s_1, t_1)) = \psi(\rho^*(s_2, t_2))$. Therefore, ψ is well-defined. Now, we show that ψ is a homomorphism. Suppose that $\rho^*(s_1, t_1)$ and $\rho^*(s_2, t_2)$ are two arbitrary elements of $(H \times T)/\rho^*$. Then,

$$\begin{aligned}
\psi(\rho^*(s_1, t_1)\blacktriangledown\rho^*(s_2, t_2)) &= \psi(\rho^*(s, t)), \text{ for all } (s, t) \in (s_1, t_1) \star (s_2, t_2) \\
&= (\rho_1^*(s), \rho_2^*(t)), \text{ for all } s \in s_1 \circ s_2,\ t \in t_1 \diamond t_2 \\
&= (\rho_1^*(s_1) \odot \rho_1^*(s_2), \rho_2^*(t_1) \otimes \rho_2^*(t_2)) \\
&= (\rho_1^*(s_1), \rho_2^*(t_1)) \times (\rho_1^*(s_2), \rho_2^*(t_2)) \\
&= \psi(\rho^*(s_1, t_1)) \times \psi(\rho^*(s_2, t_2)).
\end{aligned}$$

So, the first condition of the definition of homomorphism is verified. Now, suppose that $\rho^*(s_1, t_1) \preceq \rho^*(s_2, t_2)$. Then, $(s_1, t_1)\rho(s_2, t_2)$, which implies that $s_1\rho_1 s_2$ and $t_1\rho_2 t_2$. Thus, $\rho_1^*(s_1) \preceq_H \rho_1^*(s_2)$ and $\rho_2^*(t_1) \preceq_T \rho_2^*(t_2)$. Hence, $(\rho_1^*(s_1), \rho_2^*(t_1)) \preceq_{H\times T} (\rho_1^*(s_2), \rho_2^*(t_2))$. This means that $\psi(\rho^*(s_1, t_1)) \preceq_{H\times T} \psi(\rho^*(s_2, t_2))$, and so the second condition of the definition of homomorphism is verified. Therefore, ψ is a homomorphism. Clearly, ψ is onto. So, we show that it is one to one. Suppose that $\psi(\rho^*(s_1, t_1)) = \psi(\rho^*(s_2, t_2))$. Then, $(\rho_1^*(s_1), \rho_2^*(t_1)) = (\rho_1^*(s_2), \rho_2^*(t_2))$, and so $\rho_1^*(s_1) = \rho_1^*(s_2)$ and $\rho_2^*(t_1) = \rho_2^*(t_2)$. Hence, $(s_1, s_2) \in \rho_1^*$ and $(t_1, t_2) \in \rho_2^*$. This implies that $s_1\rho_1 s_2$, $s_2\rho_1 s_1$, $t_1\rho_2 t_2$, and $t_2\rho_2 t_1$. Thus, $(s_1, t_1)\rho(s_2, t_2)$ and $(s_2, t_2)\rho(s_1, t_1)$. Therefore, $(s_1, t_1)\rho^*(s_2, t_2)$ or $\rho^*(s_1, t_1) = \rho^*(s_2, t_2)$. Therefore, ψ is an isomorphism and the proof is completed. □

Question. Is there a regular relation ρ on an ordered semihypergroup H for which H/ρ is an ordered semihypergroup?

Gu and Tang [101] constructed an ordered regular equivalence relation on an ordered semihypergroup by hyperideals such that the corresponding quotient structure is also an ordered semihypergroup, which answers the above question. The reference of the rest of this section is [101].

Definition 3.4.15. Let $(H, \circ, \leq)$ be an ordered semihypergroup and ρ a regular (respectively strongly regular) equivalence relation on H. Then, ρ is called *ordered regular* (respectively *strongly ordered regular*) if there exists an order relation $\preceq$ on H/ρ such that:

(1) $(H/\rho, \odot, \preceq)$ is an ordered semihypergroup (respectively, semigroup);

(2) The mapping $f : H \to H/\rho$ such that $x \mapsto \rho(x)$ is a homomorphism of ordered semihypergroups.

Let H be an ordered semihypergroup and I be a hyperideal of H. We define a relation

$$\rho_I := (I \times I) \cup \{(x, y) \in H \setminus I \times H \setminus I \mid x = y\}$$

on H. It is clear that ρ_I is an equivalence relation on H. Moreover,

$$\rho_I(a) = \begin{cases} \{a\} & \text{if } a \in H \setminus I \\ I & \text{if } a \in I. \end{cases}$$

Therefore, $H/\rho_I = \{\{x\} \mid x \in H \setminus I\} \cup \{I\}$.

Theorem 3.4.16. *Let $(H, \circ, \leq)$ be an ordered semihypergroup and I be a hyperideal of H. Then, ρ_I is an ordered regular equivalence relation on H.*

Proof

(1) We first show that ρ_I is regular. If $a, b \in H$ and $a\rho_I b$, then $a, b \in I$ or $a = b \in H \setminus I$.

  (i) If $a, b \in I$, then $a \circ x \subseteq I$ and $b \circ x \subseteq I$ for any $x \in H$. Thus, for any $m \in a \circ x$, $n \in b \circ x$, we have $(m, n) \in I \times I \subseteq \rho_I$. Hence, $a \circ x \rho_I b \circ x$.

  (ii) If $a = b \in H \setminus I$, then $a \circ x = b \circ x$ for any $x \in H$. Therefore, $a \circ x \rho_I b \circ x$.

In the same way, we have $x \circ a \rho_I x \circ b$ for any $x \in H$. Hence, ρ_I is a regular equivalence relation on H. Therefore, $(H/\rho_I, \odot_I)$ is a semihypergroup, where the operation $\odot_I$ is defined by $\rho_I(a) \odot_I \rho_I(b) = \{\rho_I(c) \mid c \in a \circ b\}$ for any $a, b \in H$.

(2) We define a relation $\preceq_I$ on H/ρ_I as follows:

$$\preceq_I := \{(I, I)\} \cup \{(I, \{x\}) \mid x \in H \setminus I\} \cup \{(\{x\}, \{y\}) | x, y \in H \setminus I, x \leq y\}.$$

We show that $\preceq_I$ is an order relation.

(i) Let $\rho_I(x) \in H/\rho_I$. Then, $\rho_I(x) = \{x\}$, $x \in H \setminus I$ or $\rho_I(x) = I$. If $\rho_I(x) = \{x\}$, $x \in H \setminus I$, then $\rho_I(x) \preceq_I \rho_I(x)$ since $x \leq x$. If $\rho_I(x) = I$, then $\rho_I(x) \preceq_I \rho_I(x)$ since $I \preceq_I I$.

(ii) Let $\rho_I(x) \preceq_I \rho_I(y)$ and $\rho_I(y) \preceq_I \rho_I(x)$. If $\rho_I(x) = \{x\}$, $x \in H \setminus I$, then $\rho_I(y) = \{y\}$, $y \in H \setminus I$ and $x \leq y$ since $\rho_I(x) \preceq_I \rho_I(y)$. Moreover, $y \leq x$ since $\rho_I(y) \preceq_I \rho_I(x)$. Thus, $x = y$ and so $\rho_I(x) = \rho_I(y)$. If $\rho_I(x) = I$, then $\rho_I(y) = I$ since $\rho_I(y) \preceq_I \rho_I(x)$. Therefore, $\rho_I(x) = \rho_I(y)$.

(iii) Let $\rho_I(x) \preceq_I \rho_I(y)$ and $\rho_I(y) \preceq_I \rho_I(z)$. If $\rho_I(x) = \{x\}$, $x \in S \setminus I$, then $\rho_I(y) = \{y\}$, $y \in S \setminus I$ and $x \leq y$ since $\rho_I(x) \preceq_I \rho_I(y)$. Moreover, $\rho_I(z) = \{z\}$, $z \in S \setminus I$ and $y \leq z$ since $\rho_I(y) \preceq_I \rho_I(z)$. Hence, $x \leq z$ and thus, $\rho_I(x) \preceq_I \rho_I(z)$. If $\rho_I(x) = I$, then $\rho_I(x) \preceq_I \rho_I(z)$ since I is the minimum element in H/ρ_I.

(3) We show that $\preceq_I$ is compatible with $\odot_I$. Let $\rho_I(x)$, $\rho_I(y)$, $\rho_I(z) \in H/\rho_I$, and $\rho_I(x) \preceq_I \rho_I(y)$.

(i) If $\rho_I(x) = I$, then $\rho_I(x) \odot_I \rho_I(z) = \{\rho_I(a) | a \in x \circ z\} = \{I\}$ since $x \circ z \subseteq I$. Moreover, $I \preceq_I \rho_I(b)$ for any $\rho_I(b) \in H/\rho_I$. Hence,

$$\rho_I(x) \odot_I \rho_I(z) \preceq_I \rho_I(y) \odot_I \rho_I(z).$$

(ii) If $\rho_I(x) = \{x\}$, $x \in H \setminus I$, then $\rho_I(y) = \{y\}$, $y \in H \setminus I$ and $x \leq y$ since $\rho_I(x) \preceq_I \rho_I(y)$. Thus, $x \circ z \leq y \circ z$. If $x \circ z \subseteq I$, then $\rho_I(x) \odot_I \rho_I(z) = \{I\}$. Hence, $\rho_I(x) \odot_I \rho_I(z) \preceq_I \rho_I(y) \odot_I \rho_I(z)$. If $(x \circ z) \cap H \setminus I \neq \emptyset$, then $(y \circ z) \cap H \setminus I \neq \emptyset$. In fact, if $y \circ z \subseteq I$, then $x \circ z \subseteq I$ since $x \circ z \leq y \circ z$. This is a contradiction. For any $\rho_I(a) \in \rho_I(x) \odot_I \rho_I(z)$, we have $\rho_I(a) = I$ or $\rho_I(a) = \{a\}$, $a \in (x \circ z) \setminus I$. If $\rho_I(a) = I$, then $\rho_I(a) \preceq_I \rho_I(b)$ for any $\rho_I(b) \in \rho_I(y) \odot_I \rho_I(z)$. If $\rho_I(a) = \{a\}$, $a \in (x \circ z) \setminus I$, then there exists $b \in y \circ z$ such that $a \leq b$. Moreover, $b \notin I$. Otherwise, $a \in I$, which is a contradiction. Thus, $\rho_I(a) = \{a\} \preceq_I \{b\} = \rho_I(b)$. It follows that $\rho_I(x) \odot_I \rho_I(z) \preceq_I \rho_I(y) \odot_I \rho_I(z)$.

In the same way, we obtain that $\rho_I(z) \odot_I \rho_I(x) \preceq_I \rho_I(z) \odot_I \rho_I(y)$. Thus, we know that $(H/\rho_I, \odot_I, \preceq_I)$ is an ordered semihypergroup. Next we show that the mapping $f : H \to H/\rho_I$ defined by $f(x) = \rho_I(x)$ for any $x \in H$ is a homomorphism of ordered semihypergroups. If $x, y \in H$, then $f(x \circ y) = f(x) \odot_I f(y)$ from the definition of $\odot_I$. Let $x \leq y$. If $x \in I$, then $\rho_I(x) = I \preceq_I \rho_I(y)$. If $x \in H \setminus I$, then

$y \in H \setminus I$. Otherwise, $y \in I$ implies that $x \in I$, which is impossible. Thus, $\rho_I(x) = \{x\} \preceq_I \{y\} = \rho_I(y)$. Therefore, ρ_I is an ordered regular equivalence relation on H.

□

Example 59. Let $H = \{a, b, c, d\}$ with the hyperoperation $\circ$ and the order relation $\leq$ below:

$\circ$	a	b	c	d
a	$\{a, d\}$	$\{a, d\}$	$\{a, d\}$	a
b	$\{a, d\}$	b	$\{a, d\}$	$\{a, d\}$
c	$\{a, d\}$	$\{a, d\}$	c	$\{a, d\}$
d	a	$\{a, d\}$	$\{a, d\}$	d

$$\leq := \{(a, a), (a, b), (a, c), (b, b), (c, c), (d, b), (d, c), (d, d)\}.$$

Then, $(H, \circ, \leq)$ is an ordered semihypergroup. Let $I = \{a, d\}$. It is easy to check that I is a hyperideal of H. Then, ρ_I is a regular equivalence relation on H and $H/\rho_I = \{I, \{b\}, \{c\}\}$. Following the definitions of the hyperoperation $\odot_I$ and the order relation $\preceq_I$ on H/ρ_I in Theorem 3.4.16, we can show that $(H/\rho_I, \odot_I, \preceq_I)$ is an ordered semihypergroup. Moreover, the mapping $f : H \to H/\rho_I$ such that $x \mapsto \rho_I(x)$ is a homomorphism. Therefore, ρ_I is ordered regular.

Theorem 3.4.17. *Let I be a hyperideal of an ordered semihypergroup $(H, \circ, \leq)$. Let $\mathcal{A}$ be the set of hyperideals of H containing I and $\mathcal{B}$ be the set of hyperideals of $(H/\rho_I, \odot_I, \preceq_I)$, where the hyperoperation $\odot_I$ and the order relation $\preceq_I$ are defined as above. Then the mapping $f : \mathcal{A} \to \mathcal{B}$ such that $J \mapsto \rho_I(J) := \{\rho_I(x) | x \in J\}$ is an inclusion-preserving bijection from $\mathcal{A}$ onto $\mathcal{B}$.*

Proof

(1) We first show that $\rho_I(J)$ is a hyperideal of H/ρ_I for any $J \in \mathcal{A}$.
 - **(i)** Let $\rho_I(x) \in H/\rho_I$ and $\rho_I(y) \in \rho_I(J)$. Then, $y \in J$ and so $x \circ y \subseteq J$. Thus, $\rho_I(x) \odot_I \rho_I(y) = \{\rho_I(a) \,|a \in x \circ y\} \subseteq \rho_I(J)$. Hence, $H/\rho_I \odot_I \rho_I(J) \subseteq \rho_I(J)$. Similarly, we have $\rho_I(J) \odot_I H/\rho_I \subseteq \rho_I(J)$.
 - **(ii)** Let $\rho_I(x) \in H/\rho_I$, $\rho_I(y) \in \rho_I(J)$ and $\rho_I(x) \preceq_I \rho_I(y)$. Then, $y \in J$. If $x \in I$, then $x \in J$ and so $\rho_I(x) \in \rho_I(J)$. If $x \in H \setminus I$, then $\rho_I(x) = \{x\} \preceq_I \rho_I(y)$. Thus, $\rho_I(y) = \{y\}$ and $x \leq y$. Therefore, $x \in J$ and so $\rho_I(x) \in \rho_I(J)$.

(2) We show that θ is injective. Let $J_1, J_2 \in \mathcal{A}$ and $\rho_I(J_1) = \rho_I(J_2)$. For any $j_1 \in J_1$, either $j_1 \in I$ or $j_1 \in J_1 \setminus I$. If $j_1 \in I$, then $j_1 \in J_2$. If $j_1 \in J_1 \setminus I$, then $\rho_I(j_1) = \{j_1\} \in \rho_I(J_2)$. Thus, there exists $j_2 \in J_2$ such that $\{j_1\} = \{j_2\} = \rho_I(j_2)$. Therefore, $j_1 = j_2 \in J_2$ and so $J_1 \subseteq J_2$. By symmetry, we have $J_2 \subseteq J_1$.

(3) We show that θ is surjective. Let K be a hyperideal of H/ρ_I and $J = \{x \in H | \rho_I(x) \in K\}$. For any $a \in H$, $x \in J$, we have $\{\rho_I(b) | b \in a \circ x\} = \rho_I(a) \circ_I \rho_I(x) \subseteq K$. Thus, $a \circ x \subseteq J$, that is, $H \circ J \subseteq J$. Similarly, $J \circ H \subseteq J$. If $y \in H$ and $y \leq x \in J$, then $\rho_I(y) \preceq_I \rho_I(x) \in K$. Hence, $\rho_I(y) \in K$ and so $y \in J$. Let $x \in I$. Then, $\rho_I(x) = I$. Thus, $\rho_I(x) \preceq_I \rho_I(y) \in K$ for every $y \in J$. Hence, $\rho_I(x) \in K$ and so $x \in J$. It follows that J is a hyperideal of H containing I. Clearly, $K = \rho_I(J)$.

(4) It is obvious that θ is inclusion-preserving.

□

Theorem 3.4.18. *Let I, J be hyperideals of an ordered semihypergroup $(H, \circ, \leq)$ such that $I \subseteq J$. Then, $J/\rho_I := \{\{x\} \,|x \in J \setminus I\} \cup \{I\}$ is a hyperideal of H/ρ_I. Moreover, $(H/\rho_I)/(\rho_J/\rho_I)$ is isomorphic to H/ρ_J as ordered semihypergroups.*

Proof. It is easy to see that $J/\rho_I = \rho_I(J)$. Thus, from the proof of Theorem 3.4.17, we know that J/ρ_I is a hyperideal of H/ρ_I. By Theorem 3.4.16, we can define similarly the hyperoperations $\odot_J$, $\odot_{J/\rho_I}$ and the order relations $\preceq_J$, $\preceq_{J/\rho_I}$ on H/ρ_J and $(H/\rho_I)/(\rho_J/\rho_I)$, respectively. By the proof of Theorem 3.4.16, we have that $(H/\rho_I, \odot_I, \preceq_I)$, $(H/\rho_J, \odot_J, \preceq_J)$, and $((H/\rho_I)/(\rho_J/\rho_I), \odot_{J/\rho_I}, \preceq_{J/\rho_I})$ are ordered semihypergroups.

Consider the mapping $\phi : H/\rho_J \to (H/\rho_I)/(\rho_J/\rho_I)$ defined by

$$\phi(a) = \begin{cases} \{\{x\}\} & \text{if } a = \{x\} \text{ for some } x \in H \setminus J \\ J/\rho_I & \text{if } a = J. \end{cases}$$

Since $(H/\rho_I)/(\rho_J/\rho_I) = \{\{\{x\}\} \,|x \in H \setminus J\} \cup \{J/\rho_I\}$, it is a routine matter to show that ϕ is a bijection. By the definitions of the hyperoperations $\odot_I$, $\odot_J$, and $\odot_{J/\rho_I}$, we have $\phi(a \odot_J b) = \phi(a) \odot_{J/\rho_I} \phi(b)$ for any $a, b \in H/\rho_J$. Furthermore, for any $x, y \in H \setminus J$, we have

$$\begin{aligned} \{x\} \preceq_J \{y\} &\Leftrightarrow x \leq y \\ &\Leftrightarrow \{x\} \preceq_I \{y\} \\ &\Leftrightarrow \{\{x\}\} \preceq_{J/\rho_I} \{\{y\}\} \end{aligned}$$

and $J \preceq_J \{x\}, J/\rho_I \preceq_{J/\rho_I} \{\{x\}\}$. Therefore, ϕ is an isomorphism. □

Theorem 3.4.19. *Let I, J be two hyperideals of an ordered semihypergroup $(H, \circ, \leq)$. Then, $(I \cup J)/\rho_J \cong I/\rho_{I \cap J}$ as ordered semihypergroups.*

Proof. It is easy to see that $I \cup J$ and $I \cap J$ are hyperideals of H. Thus, by Theorem 3.4.16, we have that $((I \cup J)/\rho_J, \odot_J, \preceq_J)$ and $(I/\rho_{I\cap J}, \odot_{I\cap J}, \preceq_{I\cap J})$ are ordered semihypergroups. Moreover, $(I \cup J)/\rho_J = \{\{x\} | x \in I \setminus J\} \cup \{J\}$ and $I/\rho_{I\cap J} = \{\{x\} | x \in I \setminus J\} \cup \{I \cap J\}$. Define a mapping $\varphi : (I \cup J)/\rho_J \to I/\rho_{I\cap J}$ as follows:

$$\varphi(a) = \begin{cases} I \cap J & \text{if } a = J \\ a & \text{otherwise.} \end{cases}$$

It is obvious that φ is a bijection. Furthermore, by the definitions of the hyperoperations on ordered semihypergroups $(I \cup J)/\rho_J$ and $I/\rho_{I\cap J}$, we can easily show that φ satisfies the first condition of a homomorphism. By the definitions of the order relations on ordered semihypergroups $(I \cup J)/\rho_J$ and $I/\rho_{I\cap J}$, we can obtain that φ is an isomorphism. □

CHAPTER 4

Fundamental Relations

4.1 THE β RELATION

The main tools connecting the class of hyperstructures with the classical algebraic structures are the fundamental relations. The fundamental relation has an important role in the study of semihypergroups and especially of hypergroups.

Definition 4.1.1. Let $(H, \circ)$ be a semihypergroup and $n > 1$ be a natural number. We define the relation β_n on H as follows:

$$x\beta_n y \text{ if there exists } a_1, a_2, \ldots, a_n \text{ in } H, \text{ such that } \{x, y\} \subseteq \prod_{i=1}^{n} a_i,$$

and let $\beta = \bigcup_{n\geq 1} \beta_n$, where $\beta_1 = \{(x, x) | x \in H\}$ is the diagonal relation on H.

This relation was introduced by Koskas [102] and studied mainly by Corsini [69], Davvaz [23], Davvaz and Leoreanu-Fotea [72], Freni [103], Vougiouklis [84], and many others. Clearly, the relation β is reflexive and symmetric.

Example 60. If $(H, \circ)$ is a semihypergroup of size 2, then the relation β is transitive.

In general, denote by β^* the transitive closure of β.

Remark 17. A relation ρ^* is the *transitive closure* of a relation ρ if and only if

(1) ρ^* is transitive;

(2) $\rho \subseteq \rho^*$;

(3) for any relation θ, if $\rho \subseteq \theta$ and θ is transitive, then $\rho^* \subseteq \theta$, ie, ρ^* is the smallest relation that satisfies (1) and (2).

Remark 18. For any relation ρ, the transitive closure of ρ always exists.

Theorem 4.1.2. *β^* is the smallest strongly regular relation on H.*

Proof. We show that:

(1) β^* is a strongly regular relation on H;

Semihypergroup Theory
http://dx.doi.org/10.1016/B978-0-12-809815-8.00004-8

(2) if ρ is a strongly regular relation on H, then $\beta^* \subseteq \rho$.

(1) Let $a\beta^* b$ and x be an arbitrary element of H. It follows that there exist $x_0 = a, x_1, \ldots, x_n = b$ such that for all $i \in \{0, 1, \ldots, n-1\}$ we have $x_i \beta x_{i+1}$. Let $u_1 \in a \circ x$ and $u_2 \in b \circ x$. We check that $u_1 \beta^* u_2$. From $x_i \beta x_{i+1}$, it follows that there exists a hyperproduct P_i, such that $\{x_i, x_{i+1}\} \subseteq P_i$ and so $x_i \circ x \subseteq P_i \circ x$ and $x_{i+1} \circ x \subseteq P_i \circ x$, which means that $x_i \circ x \overline{\overline{\beta}} x_{i+1} \circ x$. Hence, for all $i \in \{0, 1, \ldots, n-1\}$ and for all $s_i \in x_i \circ x$ we have $s_i \beta s_{i+1}$. If we consider $s_0 = u_1$ and $s_n = u_2$, then we obtain $u_1 \beta^* u_2$. Then β^* is strongly regular on the right and, similarly, it is strongly regular on the left.

(2) We have $\beta_1 = \{(x, x) \mid x \in H\} \subseteq \rho$, since ρ is reflexive. Suppose that $\beta_{n-1} \subseteq \rho$ and show that $\beta_n \subseteq \rho$. If $a\beta_n b$, then there exist $x_1, \ldots, x_n$ in H, such that $\{a, b\} \subseteq \prod_{i=1}^{n} x_i$. Hence, there exists u, v in $\prod_{i=1}^{n-1} x_i$, such that $a \in u \circ x_n$ and $b \in v \circ x_n$. We have $u\beta_{n-1}v$ and, according to the hypothesis, we obtain $u\rho v$. Since ρ is strongly regular, it follows that $a\rho b$. Hence, $\beta_n \subseteq \rho$. By induction, it follows that $\beta \subseteq \rho$, whence $\beta^* \subseteq \rho$. □

Hence, the relation β^* is the smallest equivalence relation on H, such that the quotient H/β^* is a semigroup.

Definition 4.1.3. β^* is called the *fundamental relation* on H and H/β^* is called the *fundamental semigroup*.

Therefore, H/β^* is a semigroup with respect to an operation defined by

$$\beta^*(x) \odot \beta^*(y) = \beta^*(z), \text{ where } x, y \in H \text{ and } z \in x \circ y.$$

The β^* plays a critical role in semihypergroup theory; it permits the defining of a covariant functor between the category of semihypergroups and the category of semigroups. For this reason, the relation β^* has been an object of intensive study. A remarkable result concerning this relation in hypergroups was established by Freni [103]. He proved that the relation β is transitive in any hypergroup, ie, in hypergroups $\beta = \beta^*$.

There exist semihypergroups in which the relation β is not transitive.

Example 61. Let $H = \{a, b, c, d\}$ be a semihypergroup with the following hyperoperation:

$\circ$	a	b	c	d
a	$\{b, c\}$	$\{b, d\}$	$\{b, d\}$	$\{b, d\}$
b	$\{b, d\}$	$\{b, d\}$	$\{b, d\}$	$\{b, d\}$
c	$\{b, d\}$	$\{b, d\}$	$\{b, d\}$	$\{b, d\}$
d	$\{b, d\}$	$\{b, d\}$	$\{b, d\}$	$\{b, d\}$

Then, it is easy to see that $c\beta^* d$ but not $c\beta d$.

We recall the following example from [54]:

Example 62. Apart from isomorphism, there exist 10 simple semihypergroups having order 3, where the relation β is not transitive. These semihypergroups are H_1, H_2, H_3, H_4, H_5, and their respective transposed semihypergroups H_i^T, for all $i \in \{1, 2, 3, 4, 5\}$.

H_1 :

$\circ_1$	a	b	c
a	a	$\{a, b\}$	$\{a, c\}$
b	a	$\{a, b\}$	$\{a, c\}$
c	a	$\{a, b\}$	$\{a, c\}$

H_2 :

$\circ_2$	a	b	c
a	a	$\{a, b\}$	$\{a, c\}$
b	a	$\{a, b\}$	$\{a, c\}$
c	a	b	c

H_3 :

$\circ_3$	a	b	c
a	a	$\{a, b\}$	$\{a, c\}$
b	a	$\{a, b\}$	$\{a, c\}$
c	a	$\{a, b\}$	c

H_4 :

$\circ_4$	a	b	c
a	a	$\{a, b\}$	$\{a, c\}$
b	a	b	c
c	a	b	c

$H5$:

$\circ_5$	a	b	c
a	a	$\{a, b\}$	$\{a, c\}$
b	a	b	$\{a, c\}$
c	a	$\{a, b\}$	c

4.2 COMPLETE PARTS

Complete parts were introduced and studied for the first time by Koskas [102]. Later, this topic was analyzed by Corsini [69] and Sureau [104] mostly in the general theory of hypergroups. De Salvo studied complete parts from a combinatorial point of view. A generalization of them, called n-complete parts, was introduced by Migliorato [105]. Other hypergroupists gave a contribution to the study of complete parts and of the heart of a hypergroup. Among them, V. Leoreanu analyzed the structure of the heart of a hypergroup in her Ph.D. thesis.

We now present the definitions.

Definition 4.2.1. Let $(H, \circ)$ be a semihypergroup and A be a nonempty subset of H. We say that A is a *complete part* of H if for any nonzero natural number n and for all $a_1, \ldots, a_n$ of H, the following implication holds:

$$A \cap \prod_{i=1}^{n} a_i \neq \emptyset \Rightarrow \prod_{i=1}^{n} a_i \subseteq A.$$

Theorem 4.2.2. *If $(H, \circ)$ is a semihypergroup and ρ is a strongly regular relation on H, then for all z of H, the equivalence class of z is a complete part of H.*

Proof. Let $a_1, \ldots, a_n$ be elements of H, such that

$$\bar{z} \cap \prod_{i=1}^{n} a_i \neq \emptyset.$$

Then there exists $y \in \prod_{i=1}^{n} a_i$, such that $y\rho z$. The homomorphism $\pi_H : H \to H/\rho$ is good and H/ρ is a semigroup. It follows that $\pi_H(y) = \pi_H(z) = \pi_H(\prod_{i=1}^{n} a_i) = \prod_{i=1}^{n} \pi_H(a_i)$. This means that $\prod_{i=1}^{n} a_i \subseteq \bar{z}$. □

Definition 4.2.3. Let A be a nonempty part of H. The intersection of the parts of H that are complete and contain A is called the *complete closure* of A in H; it will be denoted by $\mathcal{C}(A)$.

Theorem 4.2.4. *Let $(H, \circ)$ be a semihypergroup. The following conditions are equivalent:*

(1) *For all $x, y \in H$ and $a \in x \circ y$, $\mathcal{C}(a) = x \circ y$;*

(2) *For all $x, y \in H$, $\mathcal{C}(x \circ y) = x \circ y$.*

Proof. $(1 \Rightarrow 2)$: We have

$$\mathcal{C}(x \circ y) = \bigcup_{a \in x \circ y} \mathcal{C}(a) = x \circ y.$$

$(2 \Rightarrow 1)$: From $a \in x \circ y$, we obtain $\mathcal{C}(a) \subseteq \mathcal{C}(x \circ y) = x \circ y$. This means that $\mathcal{C}(a) \cap x \circ y \neq \emptyset$, whence $x \circ y \subseteq \mathcal{C}(a)$. Therefore, $\mathcal{C}(a) = x \circ y$. □

Definition 4.2.5. A semihypergroup is *complete* if it satisfies one of the above equivalent conditions. A hypergroup is *complete* if it is a complete semihypergroup.

Lemma 4.2.6. *The following properties hold:*

(1) $A \subseteq \mathcal{C}(A)$;

(2) $\mathcal{C}(\mathcal{C}(A)) = \mathcal{C}(A)$;

(3) $A \subseteq A'$ *implies* $\mathcal{C}(A) \subseteq \mathcal{C}(A')$;

(4) *If $B \subseteq A$ and A is a complete part of H, then $\mathcal{C}(B) \subseteq A$.*

Proof. It is straightforward. □

Denote $K_1(A) = A$ and for all $n \geq 1$, denote

$$K_{n+1}(A) = \left\{ x \in H \mid \exists p \in \mathbb{N}, \exists (h_1, \ldots, h_p) \in H^p : x \in \prod_{i=1}^{p} h_i, K_n(A) \cap \prod_{i=1}^{p} h_i \neq \emptyset \right\}.$$

Then, $(K_n(A))_{n \geq 1}$ is an increasing chain of subsets of H.

Let $K(A) = \bigcup_{n\geq 1} K_n(A)$.

Theorem 4.2.7. *We have* $\mathcal{C}(A) = K(A)$.

Proof. Notice that $K(A)$ is a complete part of H. Indeed, if we suppose $K(A) \cap \prod_{i=1}^{p} x_i \neq \emptyset$, then there exists $n \geq 1$, such that $K_n(A) \cap \prod_{i=1}^{p} x_i \neq \emptyset$, which means that $\prod_{i=1}^{p} x_i \subseteq K_{n+1}(A)$.

Now, if $A \subseteq B$ and B is a complete part of H, then we show that $K(A) \subseteq B$. We have $K_1(A) \subseteq B$ and suppose $K_n(A) \subseteq B$. We check that $K_{n+1}(A) \subseteq B$. Let $z \in K_{n+1}(A)$, which means that there exists a hyperproduct $\prod_{i=1}^{p} x_i$, such that $z \in \prod_{i=1}^{p} x_i$ and $K_n(A) \cap \prod_{i=1}^{p} x_i \neq \emptyset$. Hence, $B \cap \prod_{i=1}^{p} x_i \neq \emptyset$, whence $\prod_{i=1}^{p} x_i \subseteq B$. We obtain $z \in B$. Therefore, $\mathcal{C}(A) = K(A)$. □

If $x \in H$, we denote $K_n(\{x\}) = K_n(x)$. This implies that

$$K_n(A) = \bigcup_{a \in A} K_n(a).$$

Theorem 4.2.8. *If x is an arbitrary element of a semihypergroup $(H, \circ)$, then*

(1) *for all $n \geq 2$ we have $K_n(K_2(x)) = K_{n+1}(x)$;*

(2) *the next equivalence holds: $x \in K_n(y) \Leftrightarrow y \in K_n(x)$.*

Proof

(1) We check the equality by induction. We have

$$K_2(K_2(x)) = \left\{ z{\in}H | \exists q{\in}\mathbb{N}, \exists(a_1, \ldots, a_q){\in}H^q : z{\in}\prod_{i=1}^{q} a_i, \right.$$
$$\left. K_2(x) \cap \prod_{i=1}^{q} a_i{\neq}\emptyset \right\} = K_3(x).$$

Suppose that $K_{n-1}(K_2(x)) = K_n(x)$. Then,

$$K_n(K_2(x)) = \left\{ z \in H | \exists q \in \mathbb{N}, \exists(a_1, \ldots, a_q){\in}H^q : z{\in}\prod_{i=1}^{q} a_i, \right.$$
$$\left. K_{n-1}(K_2(x)) \cap \prod_{i=1}^{q} a_i \neq \emptyset \right\} = K_{n+1}(x).$$

(2) We check the equivalence by induction. For $n = 2$, we have

$$x \in K_2(y) = \left\{ z \in H | \exists q \in \mathbb{N}, \exists (a_1, \ldots, a_q) \in H^q : z \in \prod_{i=1}^{q} a_i, \right.$$
$$\left. K_1(y) \cap \prod_{i=1}^{q} a_i \neq \emptyset \right\}.$$

Hence, $\{y, x\} \subseteq \prod_{i=1}^{q} a_i$, whence $y \in K_2(x)$.
Suppose that the following equivalence holds:

$$x \in K_{n-1}(y) \Leftrightarrow y \in K_{n-1}(x),$$

and we check $x \in K_n(y) \Leftrightarrow y \in K_n(x)$. If $x \in K_n(y)$, then there exists $\prod_{i=1}^{p} a_i$ with $x \in \prod_{i=1}^{p} a_i$ and there exists $v \in \prod_{i=1}^{p} a_i \cap K_{n-1}(y)$. It follows that $v \in K_2(x)$ and $y \in K_{n-1}(v)$. Hence, $y \in K_{n-1}(K_2(x)) = K_n(x)$. Similarly, we obtain the converse implication. □

Corollary 4.2.9. *$x \in C(y)$ if and only if $y \in C(x)$.*

Corollary 4.2.10. *The binary relation defined as*

$$xKy \Leftrightarrow \exists n \geq 1, \quad x \in K_n(y)$$

is an equivalence relation.

Theorem 4.2.11. *The equivalence relations K and β^* coincide.*

Proof. If $x\beta y$, then x and y belong to the same hyperproduct and so, $x \in K_2(y) \subseteq K(y)$. Hence, $\beta \subseteq K$, whence $\beta^* \subseteq K$. Now, if we have xKy and $x \neq y$, then there exists $n \geq 1$, such that $xK_{n+1}y$, which means that there exists a hyperproduct P_1, such that $x \in P_1$ and $P_1 \cap K_n(y) \neq \emptyset$. Let $x_1 \in P_1 \cap K_n(y)$. Hence, $x\beta x_1$. From $x_1 \in K_n(y)$, it follows that there exists a hyperproduct P_2, such that $x_1 \in P_2$ and $P_2 \cap K_{n-1}(y) \neq \emptyset$. Let $x_2 \in P_2 \cap K_{n-1}(y)$. Hence, $x_1\beta x_2$ and $x_2 \in K_{n-1}(y)$. After a finite number of steps, we obtain that there exist x_{n-1}, x_n such that $x_{n-1}\beta x_n$ and $x_n \in K_{n-(n-1)}(y) = \{y\}$. Therefore, $x\beta^* y$. □

Theorem 4.2.12. *If B is a nonempty subset of H, then we have*

$$C(B) = \bigcup_{b \in B} C(b).$$

Proof. Clearly, for all $b \in B$, we have $C(b) \subseteq C(B)$. On the other hand, $C(B) = \bigcup_{n \geq 1} K_n(B)$. We shall prove by induction. For $n = 1$, we have $K_1(B) = B = \bigcup_{b \in B} K_1(b)$. Suppose that $K_n(B) \subseteq \bigcup_{b \in B} K_n(b)$. If $z \in K_{n+1}(B)$, then there exists a hyperproduct P such that $z \in P$ and $K_n(B) \cap$

$P \neq \emptyset$, whence there exists $b \in B$ such that $K_n(b) \cap P \neq \emptyset$. Hence, $z \in K_{n+1}(b)$. We obtain $K_{n+1}(B) \subseteq \bigcup_{b\in B} K_{n+1}(b)$. Therefore, $\mathcal{C}(B) = \bigcup_{b\in B} \mathcal{C}(b)$. □

Lemma 4.2.13. *Let $(H, \circ)$ be a complete semihypergroup. Then, for all $x_1, \ldots, x_n \in H$ we have*

$$\mathcal{C}\left(\prod_{i=1}^{n} x_i\right) = \mathcal{C}(z), \qquad \text{for all } z \in \prod_{i=1}^{n} x_i.$$

Proof. Suppose that $z \in \prod_{i=1}^{n} x_i$. Clearly, we have $\mathcal{C}(z) \subseteq \mathcal{C}\left(\prod_{i=1}^{n} x_i\right)$. By Theorem 4.2.12, we have

$$\mathcal{C}\left(\prod_{i=1}^{n} x_i\right) = \bigcup_{z\in\prod_{i=1}^{n} x_i} \mathcal{C}(z).$$

Let $y \in \prod_{i=1}^{n} x_i$. Then, $y\beta^* z$. This implies that $y \in \mathcal{C}(z)$. Therefore,

$$\mathcal{C}\left(\prod_{i=1}^{n} x_i\right) = \prod_{i=1}^{n} x_i \subseteq \mathcal{C}(z).$$

□

Theorem 4.2.14. *Let $(H, \circ)$ be a semihypergroup. Then, H is complete if and only if for all $x_1, \ldots, x_m, y_1, \ldots, y_n$ in H,*

$$\prod_{i=1}^{m} x_i \cap \prod_{i=1}^{n} y_i \neq \emptyset \Rightarrow \prod_{i=1}^{m} x_i = \prod_{i=1}^{n} y_i.$$

Proof. Suppose that H is complete. Then, we have

$$\prod_{i=1}^{m} x_i = \mathcal{C}\left(\prod_{i=1}^{m} x_i\right) \quad \text{and} \quad \prod_{i=1}^{n} y_i = \mathcal{C}\left(\prod_{i=1}^{n} y_i\right).$$

So, it is enough to show that

$$\mathcal{C}\left(\prod_{i=1}^{m} x_i\right) = \mathcal{C}\left(\prod_{i=1}^{n} y_i\right).$$

Since $\prod_{i=1}^{m} x_i \cap \prod_{i=1}^{n} y_i \neq \emptyset$, it follows that there exists $a \in \prod_{i=1}^{m} x_i \cap \prod_{i=1}^{n} y_i$. By Lemma 4.2.13, we have

$$\mathcal{C}(a) = \mathcal{C}\left(\prod_{i=1}^{m} x_i\right) \text{ and } \mathcal{C}(a) = \mathcal{C}\left(\prod_{i=1}^{n} y_i\right),$$

and so $\mathcal{C}\left(\prod_{i=1}^{m} x_i\right) = \mathcal{C}\left(\prod_{i=1}^{n} y_i\right)$.

The proof of the converse case is straightforward. □

Proposition 4.2.15. *Let $(H, \circ)$ be a semihypergroup and A be a cyclic part of H with generator x. Then, for every $a \in A$, there exists $m \in \mathbb{N}$ such that $x^m \subseteq C(a)$. Moreover, if H is complete, then $x^m = C(a)$.*

Proof. For every $a \in A$, there exists $m \in \mathbb{N}$ such that $a \in x^m$. Suppose that b is an arbitrary element of x^m. Then, we have $a\beta b$ and so $a\beta^* b$. Hence, $b \in C(a)$. Thus, $x^m \subseteq C(a)$.

Now, let H be complete. Since $a \in x^m$, by Lemma 4.2.13, it follows that $C(a) = C(x^m) = x^m$. □

Corollary 4.2.16. *Let $(H, \circ)$ be a complete semihypergroup. Then,*

$$a \circ b = C(a \circ b) = C(a) \circ b = a \circ C(b) = C(a) \circ C(b),$$

for all $a, b \in H$.

Proposition 4.2.17. *If $(H, \circ)$ is a cyclic and complete semihypergroup, then $(H, \circ)$ is commutative.*

Proof. Suppose that h is a generator of H. If $x, y \in H$, then there exist positive integers m, n such that $x \in h^m$ and $y \in h^n$. So, $x \circ y \subseteq h^{m+n}$, which implies that $(x \circ y) \cap h^{m+n} \neq \emptyset$. Now, by Theorem 4.2.14, we conclude that $x \circ y = h^{m+n}$. Similarly, we obtain $y \circ x = h^{m+n}$. Thus, $x \circ y = y \circ x$. □

Set $cycl(h) = \min\{n \in \mathbb{N} | n \geq 2,\ h \in h^n\}$, otherwise $cycl(h) = \infty$.

Theorem 4.2.18. *If $(H, \circ)$ is a cyclic and complete semihypergroup with generator h such that $cycl(h) = k$, then $(H, \circ)$ is a hypergroup.*

Proof. Suppose that a, b are arbitrary elements of H. We should prove the following statements:

(1) There exists $x \in H$ such that $a \in b \circ x$.

(2) There exists $x \in H$ such that $a \in y \circ b$.

We prove (1). Since H is cyclic with generator h, it follows that there exist $m, n \in \mathbb{N}$ such that

$$a \in h^m \quad \text{and} \quad b \in h^n.$$

If $m > n$, then for every $x \in h^{m-n}$ we obtain

$$h^n \circ x \subseteq h^n \circ h^{m-n} = h^m.$$

This implies that $(h^n \circ x) \cap h^m \neq \emptyset$. By Theorem 4.2.14, we conclude that $h^n \circ x = h^m$ and so $a \in h^n \circ x$. Since $b \in h^n$, by Proposition 4.2.15 we obtain $h^n = C(b)$. Thus, $a \in C(b) \circ x$. Since $C(b) \circ x = C(b \circ x) = b \circ x$, it follows that $a \in b \circ x$.

If $m \leq n$, then

$$h^m \subseteq (h^k)^m = h^{km}, \quad \text{where} k = cycl(h) \geq 2.$$

This implies that $a \in h^{km}$. Thus, we have

$$a \in h^{km} \quad \text{and} \quad b \in h^{n}.$$

If $km > n$, then similar to the above argument, the proof completes. Otherwise, we continue the above process to obtain $k^t m > n$.

The proof of statement (2) is similar to (1). □

Proposition 4.2.19. *Let $(H, \circ)$ be a semihypergroup and P be a complete part of H generated by x. Then, for all $a_1, \ldots, a_n \in P$, there exists $m \in \mathbb{N}$ such that*

$$x^m \subseteq \mathcal{C}\left(\prod_{i=1}^{n} a_i\right).$$

If H is complete, then

$$x^m = \mathcal{C}\left(\prod_{i=1}^{n} a_i\right).$$

Proof. Suppose that $a_1, \ldots, a_n \in P$. Then, there exist $m_1, \ldots, m_n \in \mathbb{N}$ such that $a_i \in x^{m_i}$ (for $i = 1, \ldots, n$). So, $\prod_{i=1}^{n} a_i \subseteq x^{m_1+\cdots+m_n}$. If we set $m = m_1 + \cdots + m_n$, then $\prod_{i=1}^{n} a_i \subseteq x^m$.

For every $y \in x^m$ and $b \in \prod_{i=1}^{n} a_i$, we obtain $b\beta^* y$. This implies that $y \in \mathcal{C}(b)$. Since $\mathcal{C}(b) \subseteq \mathcal{C}\left(\prod_{i=1}^{n} a_i\right)$, it follows that $y \in \mathcal{C}\left(\prod_{i=1}^{n} a_i\right)$. Thus, $x^m \subseteq \mathcal{C}\left(\prod_{i=1}^{n} a_i\right)$.

Now, suppose that H is complete. Since $\prod_{i=1}^{n} a_i \subseteq x^m$, it follows that $\left(\prod_{i=1}^{n} a_i\right) \cap x^m \neq \emptyset$. So, by Theorem 4.2.14, we obtain $\prod_{i=1}^{n} a_i = x^m$. Therefore, $\mathcal{C}\left(\prod_{i=1}^{n} a_i\right) = x^m$. □

Definition 4.2.20. Let $(H, \circ)$ be a semihypergroup and h be an element of H. We define the relation Δ_h on H as follows:

$$x \, \Delta_h \, y \Leftrightarrow \text{there exist } m \in \mathbb{N} \setminus \{1\} \text{ such that } \{x, y\} \subseteq h^m.$$

We denote Δ_H^* the transitive closure of Δ_h.

Theorem 4.2.21. *If $(H, \circ)$ is a cyclic semihypergroup with generator h, then $\Delta_h^* = \beta^*$.*

Proof. Suppose that $x \, \Delta_h^* \, y$. Then, there exist $z_1, \ldots, z_n \in H$ such that

$$x = z_1 \, \Delta_h^* \, z_2, \ldots, z_{n-1} \, \Delta_h^* \, z_n = y.$$

Thus, we have

$$x = z_1 \beta z_2, \ldots, z_{n-1} \beta z_n = y.$$

This implies that $x\beta^* y$.

Conversely, suppose that $x\beta^* y$. Then, there exist $z_1, \ldots, z_n \in H$ such that $x = z_1\beta z_2, \ldots, z_{n-1}\beta z_n = y$. So, there exist $x_{i_1}, \ldots, x_{i_k} \in H$ such that

$$\{z_i, z_{i-1}\} \subseteq \prod_{j=1}^{k} x_{i_j} \quad (\text{for } i = 2, \ldots, n).$$

Since each x_{i_j} contains in a power of h, it follows that

$$\{z_i, z_{i-1}\} \subseteq h^{r_i} \quad (\text{for some } r_i \in \mathbb{N}).$$

Therefore, we obtain $z_{i-1}\ \Delta_h\ z_i$ $(i = 2, \ldots, n)$. This implies that $x\ \Delta_h^*\ y$. □

Theorem 4.2.22. *If $(H, \circ)$ is a complete and cyclic semihypergroup with generator h, then $\Delta_h^* = \beta$.*

Proof. By Theorem 4.2.21, $\Delta_h \subseteq \Delta_h^* = \beta^*$. So, it is enough to show that $\beta^* \subseteq \Delta_h$. Suppose that $x\beta^* y$. Then, $x \in \mathcal{C}(y)$. Since H is complete, it follows that $\mathcal{C}(x) = \mathcal{C}(y)$. By Proposition 4.2.19, there exist $m, n \in \mathbb{N}$ such that $\mathcal{C}(x) = h^m$ and $\mathcal{C}(y) = h^n$. Thus, $x\ \Delta_h\ y$. □

Theorem 4.2.23. *Let $(H, \circ)$ and $(H', \star)$ be semihypergroups and $f : H \to H'$ be a good homomorphism. Then,*

(1) *$f(\mathcal{C}(x)) \subseteq \mathcal{C}(f(x))$, for all $x \in H$;*

(2) *f determines a semigroup homomorphism $f^* : H/K \to H'/K'$ defined by $f^*(\pi_H(x)) = \pi_{H'}(f(x))$.*

Proof

(1) It is sufficient to observe that the following implication is valid:

$$x\beta_n y \Rightarrow f(x)\beta_n' f(y).$$

(2) Suppose that $\pi_H(x) = \pi_H(y)$. Then, xKy. By (1), we obtain $f(x)K'f(y)$. Thus, $f^*(\pi_H(x)) = f^*(\pi_H(y)$. So, f^* is well defined. Moreover, for every $x, y \in H$, we have

$$\begin{aligned} f^*(\pi_H(x) \odot \pi_H(y)) &= f^*(\pi_H(u)), \text{ for all } u \in x \circ y \\ &= \pi_{H'}(f(u)) \\ &= \pi_{H'}(f(x) \star f(y)) \\ &= \pi_{H'}(f(x)) \odot' \pi_{H'}(f(y)) \\ &= f^*(\pi_H(x)) \odot' f^*(\pi_H(x)). \end{aligned}$$

Thus, f^* is a semigroup homomorphism. □

Let $(H, \circ)$ be a semihypergroup. Denote

$$P_1(H) = \{\{x\} | x \in H\};$$
$$P_n(H) = \left\{ \prod_{i=1}^{n} x_i | x_1, \ldots, x_n \in H \right\}, \text{ where } n \geq 2;$$
$$P(H) = \bigcup_{n=1}^{\infty} P_n(H).$$

Remark 19. By using the above subsets, we can consider the relation β_n as follows:

$$x\beta_n y \Leftrightarrow \text{there exists } Q \in P_n(H) \text{ such that } x, y \in Q.$$

Now, we present an example for obtaining the complete closure of some subsets of a semihypergroup.

Example 63. Let H be a set having at least four elements and let $a \in H$. Consider A and B, two subsets of H such that $a \notin A \cup B$, $A \cap B \neq \emptyset$, $A \not\subset B$, and $B \not\subset A$. Define on H a hyperoperation by

$$x \circ y = \begin{cases} B & \text{if } (x, y) = (a, a) \\ A & \text{otherwise.} \end{cases}$$

Then, for every $x, y, z \in H$, $x \circ (y \circ z) = (x \circ y) \circ z = A$. Hence, $(H, \circ)$ is a semihypergroup. In this semihypergroup, $P_2(H) = \{A, B\}$ and $P_n(H) = \{A\}$, for every $n \geq 3$. If $b \in B \setminus A$, then $K_2(b) = B$ and $K_3(b) = \mathcal{C}(b) = A \cup B$.

In the following, Theorem 4.2.26 shows that it is more simple to determine the complete closure for hypergroups than for semihypergroups.

Lemma 4.2.24. *Let $(H, \circ)$ be a hypergroup, $Q \in P(H)$ and $a \in H$. Then, there exists $Q' \in P(H)$ such that $Q \subseteq Q' \circ a$.*

Proof. It is straightforward. □

Lemma 4.2.25. *Let $(H, \circ)$ be a hypergroup and $Q, Q' \in P(H)$ with $Q \cap Q' \neq \emptyset$. Then, for every $a \in Q$, there exists $Q'' \in P(H)$ such that $Q' \cup \{a\} \subseteq Q''$.*

Proof. Suppose that $b \in Q \cap Q'$. By Lemma 4.2.24, there exists $Q_1 \in P(H)$ such that $Q' \subseteq Q_1 \circ a$. In the same way, as $b \circ H = H$, there exists $x \in H$ such that $a \in b \circ x$. Then, we have

$$Q' \subseteq Q_1 \circ a \subseteq Q_1 \circ b \circ x \subseteq Q_1 \circ Q \circ x = Q'' \in P(H)$$

and

$$Q'' \supseteq Q_1 \circ a \circ x \supseteq Q' \circ x \supseteq b \circ x \supseteq \{a\}.$$

Therefore, $Q' \cup \{a\} \subseteq Q''$ and $Q'' \in P(H)$. □

Theorem 4.2.26. *Let $(H, \circ)$ be a hypergroup and consider B a subset of H. Then, $\mathcal{C}(B) = K_2(B)$.*

Proof. It suffices to prove that for every $b \in B$, $\mathcal{C}(b) = K_2(b)$. In order to prove it, it is sufficient to establish that $K_2(b)$ is a complete part of H. Let $Q \in P(H)$ such that $Q \cap K_2(b) \neq \emptyset$. Using the definition of $K_2(b)$, we deduce that there exists $Q' \in P(H)$ such that $b \in Q'$ and $Q \cap Q' \neq \emptyset$. Then, by Lemma 4.2.25, $P(H)$ contains a subset Q'' with $Q \cup \{b\} \subseteq Q''$. Hence, we obtain $Q \subseteq K_2(b)$. □

Example 63 shows that generally $\mathcal{C}(b) \neq K_2(b)$. In fact, in Example 63, we have $\mathcal{C}(b) = K_3(b)$, for every $b \in H$.

Now, we give another example in order to show that the chain $(K_n(b))_{n\geq 1}$ can be strictly increasing.

Example 64. Let H be an infinite set. Consider a family $(a_m)_{m\geq 1}$ of distinct elements of H and a family $(A_n)_{n\geq 0}$ of nonempty subsets of H such that the following conditions hold:

(1) $A_n \cap A_{n+m} = \emptyset$, for every $m \geq 2$ and $n \geq 0$;
(2) $A_n \cap A_{n+1} \neq \emptyset$ and $A_{n+1} \not\subset A_n$, for every $n \geq 0$;
(3) $a_m \notin \bigcup_{n\geq 0} A_n$, for every $m \geq 1$;
(4) $A_0 \not\subset A_1$.

For instance, if $H = \mathbb{N} \cup \{0\}$ and

$$a_m = 2m, \quad \text{for every } m \in \mathbb{N},$$
$$A_n = \{2n+1, 2n+3\}, \quad \text{for every } n \in \mathbb{N} \cup \{0\},$$

the conditions (1)–(4) hold.

We define on H a hyperoperation by

$$x \circ y = \begin{cases} A_m & \text{if there exists } m \in \mathbb{N} \text{ such that } x = y = a_m \\ A_0 & \text{otherwise.} \end{cases}$$

Then, $x \circ (y \circ z) = (x \circ y) \circ z = A_0$, for every $x, y, z \in H$, ie, H is a semihypergroup. If $b \in A_0 \setminus A_1$, then for every $n \in \mathbb{N} \cup \{0\}$, $K_{n+2}(b) = A_0 \cup \cdots \cup A_n$. This means that, in this example, the chain $(K_n(b))_{n\geq 1}$ is strictly increasing.

4.3 THE TRANSITIVITY OF THE RELATION β IN SEMIHYPERGROUPS

It is known that in a semihypergroup $(H, \circ)$,

$$x\beta^* y \Leftrightarrow \mathcal{C}(x) = \mathcal{C}(y) \Leftrightarrow x \in \mathcal{C}(y).$$

In Example 63, $\beta_2 = A \times A \cup B \times B$, $\beta_3 = \beta_4 = \cdots = A \times A$, $\beta = \beta_1 \cup \beta_2$, and $\beta^* = [(A \cup B) \times (A \cup B)] \cup \beta_1$. Hence, in this case, β is not transitive and the chain $(\beta_n)_{n \in \mathbb{N}}$ is not increasing. In the sequel, we present a characterization for the semihypergroups in which the chain $(\beta_n)_{n \in \mathbb{N}}$ is increasing. The main references for this section are [54, 58].

Lemma 4.3.1. *Let $(H, \circ)$ be a global idempotent semihypergroup (ie, $H = H \circ H$). Then, for every $n \in \mathbb{N}$ and every $Q \in P_n(H)$, there exists $Q' \in P_{n+1}(H)$ such that $Q \subseteq Q'$.*

Proof. Suppose that $Q = x_1 \circ \cdots \circ x_n$, where $x_1, \ldots, x_n$ are in H. There exist x'_n, x''_n in H such that $x_n \in x'_n \circ x''_n$. Then, $Q \subseteq x_1 \circ \cdots \circ x_{n-1} \circ x'_n \circ x''_n = Q'$ and $Q' \in P_{n+1}(H)$. □

Theorem 4.3.2. *Let $(H, \circ)$ be a semihypergroup. The chain $(\beta_n)_{n \in \mathbb{N}}$ of relations on H is increasing if and only if H is global idempotent.*

Proof. We first remark that $\beta_1 \subseteq \beta_2$ if and only if H is global idempotent. It remains to prove that if H is a global idempotent, then $\beta_n \subseteq \beta_{n+1}$ for every $n \in \mathbb{N}$. This is an immediate consequence of the previous lemma. □

Corollary 4.3.3. *If $(H, \circ)$ is a finite global idempotent semihypergroup, then there exists $n \in \mathbb{N}$ such that $\beta = \beta_n$. In particular, if $(H, \circ)$ is a finite hypergroup, then there exists $n \in \mathbb{N}$ such that $\beta = \beta_n$.*

Now, we give an example of hypergroup for which $\beta_n \neq \beta_{n+1}$ for every $n \in \mathbb{N}$.

Example 65. Let $(G, +)$ be an abelian group and let a be an element of infinite order in G. We define on G a hyperoperation by

$$x \oplus y = \{x + y, x + y + a\}, \quad \text{for all } x, y \in G.$$

Then, $(G, \oplus)$ is a hypergroup for which

$$P_{n+1}(G) = \bigcup \{x, x + a, \ldots, x + na | x \in G\},$$

where $n \in \mathbb{N}$. Thus, in G,

$$x \beta_n y \Leftrightarrow x - y \in \{-na, \ldots, -a, 0, a, \ldots, na\},$$

that is, $\beta_n \subseteq \beta_{n+1}$, but $\beta_n \neq \beta_{n+1}$, for all $n \in \mathbb{N}$.

As we have already mentioned, if $(H, \circ)$ is a hypergroup, then H/β^* is a group. Denote by 1 its identity. Then, $\pi_H^{-1}(1) = \omega_H$ is called the *heart* of H.

For every $x \in \omega_H$, $\mathcal{C}(x) = \omega_H = K_2(x)$.

The next result is in connection with a problem raised up by Corsini and Freni [70], namely to characterize the hypergroups for which ω_H is a hyperproduct.

Theorem 4.3.4. *Let $(H, \circ)$ be a hypergroup such that the set $P(H)\setminus P_1(H)$ of all proper hyperproducts of H is finite. Then, the following properties hold:*

(1) *For every $Q \in P(H)$ and $Q' \in P(H)$, with $Q \cap Q' \neq \emptyset$, there exists $Q'' \in P(H)$ such that $Q \cup Q' \subseteq Q''$;*

(2) *For every $x \in H$, $\mathcal{C}(x) \in P(H)$;*

(3) $\omega_H \in P(H)$.

Proof

(1) If $Q \subseteq Q'$ we can choose $Q'' = Q'$. Otherwise, let $x \in Q \setminus Q'$. By Lemma 4.2.25, we get that there exists $Q'_1 \in P(H)$ such that $Q' \cup \{x\} \subseteq Q'_1$. Hence, $Q' \subseteq Q'_1$ and $Q' \neq Q'_1$. But Q' is strictly contained only in a finite number of hyperproducts of H. Thus, with an analogous reasoning, after a finite number of steps, we shall find $Q'' \in P(H)$ such that $Q \cup Q' \subseteq Q''$.

(2) Since $\mathcal{C}(x) = \cup\{Q \in P(H) \mid x \in Q\}$ and $P(H) \setminus P_1(H)$ is finite, it follows that there exist $n \in \mathbb{N}$ and $Q_1, \ldots, Q_n$ in $P(H)$ such that $\mathcal{C}(x) = Q_1 \cup \cdots \cup Q_n$. Now, according to (1) we deduce that $\mathcal{C}(x) \in P(H)$.

(3) It is a direct consequence of (2).

□

Corollary 4.3.5. *If $(H, \circ)$ is a finite hypergroup, then $\omega_H \in P(H)$.*

In the next example, we present a hypergroup for which the heart is not a hyperproduct.

Example 66. Let H be an uncountable set endowed with a hyperoperation defined by

$$x \circ y = \{x, y\}, \qquad \text{for all } x, y \in H.$$

Then, $(H, \circ)$ is a hypergroup such that $\beta_2 = \beta_3 = \cdots = H \times H$, $P_n(H) = \{X \subseteq H \mid 1 \leq |X| \leq n\}$, where $n \in \mathbb{N}$, $\omega_H = H$, and $P(H) = \{X \subseteq H \mid 1 \leq |X| \leq \chi_0\}$. Hence, $\omega_H \notin P(H)$.

The purpose of the next theorem is to characterize the semihypergroups for which the relation β is transitive.

Theorem 4.3.6. *Let $(H, \circ)$ be a semihypergroup. Then, the relation β is transitive in H if and only if $\mathcal{C}(x) = K_2(x)$, for all $x \in H$.*

Proof. Assume that $\mathcal{C}(x) = K_2(x)$, for all $x \in H$. Let $a, b, c \in H$ such that $a\beta b$ and $b\beta c$. Then, there exist $Q \in P(H)$ and $Q' \in P(H)$ such that $\{a\} \cup \{b\} \subseteq Q$ and $\{b\} \cup \{c\} \subseteq Q'$. Therefore, $Q' \cap \mathcal{C}(a) \neq \emptyset$. Thus, $c \in \mathcal{C}(a) = K_2(a)$. This implies that there exists $Q'' \in P(H)$ such that $\{a\} \cup \{c\} \subseteq Q''$, ie, $a\beta c$.

Now, we prove the converse. Suppose that β is transitive. We have to establish that $\mathcal{C}(x) = K_2(x)$, for all $x \in H$. For doing this, it suffices to show that $K_2(x)$ is a complete part of H, ie, if $Q \in P(H)$ and $Q \cap K_2(x) \neq \emptyset$, then $Q \subseteq K_2(x)$.

Since $Q \cap K_2(x) \neq \emptyset$, it follows that there exists $Q' \in P(H)$ such that $x \in Q'$ and $Q \cap Q' \neq \emptyset$.

Let $y \in Q \cap Q'$ and $z \in Q$. Then, $z\beta y$ and $y\beta x$, whence $z\beta x$. Thus, there exists $Q'' \in P(H)$ such that $\{z\} \cup \{x\} \subseteq Q''$. Therefore, $z \in K_2(x)$. As z is arbitrarily chosen, it follows that $Q \subseteq K_2(x)$. Hence, $K_2(x)$ is a complete part of H and $\mathcal{C}(x) = K_2(x)$, for all $x \in H$. □

By using Theorems 4.2.26 and 4.3.6, we obtain the following result:

Corollary 4.3.7. *If $(H, \circ)$ is a hypergroup, then the relation β is transitive.*

Now, we present a class of semihypergroups for which the relation β is transitive.

Theorem 4.3.8. *Let $(H, \circ)$ be a finite semihypergroup and let $Max(P(H))$ be the set of all maximal elements of the ordered set $(P(H), \subseteq)$. If $Max(P(H))$ is a partition of H, then $\mathcal{C}(x) \in P(H)$, for every $x \in H$, and the relation β is transitive on H.*

Proof. In the case of finite semihypergroups, every hyperproduct is contained in a maximal hyperproduct. On the other hand, $\cup\{Q | Q \in \mathrm{Max}(P(H))\} = H$. This implies that for all $x \in H$, $\mathcal{C}(x)$ is a union of sets of $\mathrm{Max}(P(H))$. But $\mathrm{Max}(P(H))$ is a partition of H. It follows that there exists only one set $Q \in \mathrm{Max}(P(H))$ such that $\mathcal{C}(x) = Q$. The proof may be now completed as a direct consequence of Theorem 4.3.6. □

There exist (finite) semihypergroups H in which the relation β is transitive but $\mathrm{Max}(P(H))$ is not a partition of H. This follows from the following example:

Example 67. Let H be a set having at least five elements. Consider a, b, two distinct elements of H and A, B, C three nonempty and mutually disjoint subsets of H such that a, b are not in $A \cup B \cup C$. We define on H a hyperoperation by

$$x \circ y = \begin{cases} A \cup B & \text{if } (x, y) = (a, a) \\ B \cup C & \text{if } (x, y) = (b, b) \\ C \cup A & \text{otherwise.} \end{cases}$$

Then, $x \circ (y \circ z) = (x \circ y) \circ z$, for all $x, y, z \in H$, whence H is a semihypergroup. For this semihypergroup,

$$C(x) = K_2(x) = \begin{cases} A \cup B \cup C & \text{if } x \in A \cup B \cup C \\ \{x\} & \text{if } x \in H \setminus (A \cup B \cup C). \end{cases}$$

Thus, the relation β is transitive in H. But $\text{Max}(P(H)) = \{A \cup B, A \cup C, B \cup C\} \cup \{\{x\} | x \in H \setminus (A \cup B \cup C)\}$. Hence, $\text{Max}(P(H))$ is not a partition of H.

The forthcoming theorem concerns the transitivity of fundamental relation β in a hypercyclic simple semihypergroup. First, we need the following lemma.

Lemma 4.3.9. *Let $(H, \circ)$ be a simple semihypergroup and let $P = \prod_{i=1}^{n} x_i$, $Q = \prod_{i=1}^{m} y_i$ be two hyperproducts of elements in H. If $P \cap Q \neq \emptyset$ and $z \in Q$, then there exist a hyperproduct $T = \prod_{i=1}^{k} z_i$ of elements in H and a permutation $\sigma \in \mathbb{S}_k$ such that $z \in T$ and $P \subseteq \prod_{i=1}^{k} z_{\sigma(i)}$.*

Proof. The result is obvious if $n = 1$, taking $T = Q$ and σ as the identity of $\mathbb{S}_m$. So let $n > 1$ and $b \in P \cap Q$. Since H is simple, it follows that there exist $c_1, c_2, d_1, d_2 \in H$ such that $z \in c_1 \circ b \circ c_2$ and $x_n \in d_1 \circ z \circ d_2$. Therefore, we have

$$\begin{aligned} z \in c_1 \circ b \circ c_2 \subseteq c_1 \circ P \circ c_2 &= c_1 \circ \left(\prod_{i=1}^{n-1} x_i\right) \circ x_n \circ c_2 \\ &\subseteq c_1 \circ \left(\prod_{i=1}^{n-1} x_i\right) \circ d_1 \circ z \circ d_2 \circ c_2 \\ &\subseteq c_1 \circ \left(\prod_{i=1}^{n-1} x_i\right) \circ d_1 \circ \left(\prod_{i=1}^{m} y_i\right) \circ d_2 \circ c_2. \end{aligned}$$

Moreover,

$$\begin{aligned} P = \left(\prod_{i=1}^{n} x_i\right) = \left(\prod_{i=1}^{n-1} x_i\right) \circ x_n &\subseteq \left(\prod_{i=1}^{n-1} x_i\right) \circ d_1 \circ z \circ d_2 \\ &\subseteq \left(\prod_{i=1}^{n-1} x_i\right) \circ d_1 \circ c_1 \circ b \circ c_2 \circ d_2 \\ &\subseteq \left(\prod_{i=1}^{n-1} x_i\right) \circ d_1 \circ c_1 \circ \left(\prod_{i=1}^{m} y_i\right) \circ c_2 \circ d_2. \end{aligned}$$

Finally, setting $c_1 = z_1, x_1 = z_2, \ldots, x_{n-1} = z_n, d_1 = z_{n+1}, y_1 = z_{n+2}, \ldots, y_m = z_{n+m+1}, d_2 = z_{n+m+2}, c_2 = z_{n+m+3}$, and $k = n + m + 3$, there exists a permutation $\sigma \in \mathbb{S}_k$ such that

$$z \in \prod_{i=1}^{k} z_i \quad \text{and} \quad P \subseteq \prod_{i=1}^{k} z_{\sigma(i)}.$$

□

Theorem 4.3.10. *If $(H, \circ)$ is a hypercyclic simple semihypergroup, then the relation β is transitive.*

Proof. Suppose that there exists a hyperproduct P of elements in H such that $H = \bigcup_{m \geq 1} P^m$. Let a, b, c be three elements in H such that $a\beta b$ and $b\beta c$. From hypotheses, there exist two hyperproducts A and B of elements in H such that $\{a, b\} \subseteq A$ and $\{b, c\} \subseteq B$. Since $A \cap B \neq \emptyset$ and $c \in B$, by Lemma 4.3.9, there exist a hyperproduct $\prod_{i=1}^{k} z_i$ of elements in H and a permutation $\sigma \in \mathbb{S}_k$ such that $A \subseteq \prod_{i=1}^{k} z_i$ and $c \in \prod_{i=1}^{k} z_{\sigma(i)}$. Moreover, there exist positive integers $n_1, n_2, \ldots, n_k$ such that $z_i \in P^{n_i}$, for all $i \in \{1, 2, \ldots, k\}$. Thus, setting $n = n_1 + \cdots + n_k$, we obtain $A \cup \{c\} \subseteq P^n$ and finally $a\beta c$. □

In the next proposition, we use Theorem 4.3.10 to find the maximum cardinality of hyperproducts of semihypergroups H that fulfill the following conditions:

(1) All subsemihypergroups of H (H itself included) are simple;

(2) The relation β in H and the relation β_K in all subsemihypergroups $K \subset H$ of size ≥ 3 are not transitive.

These semihypergroups are called *fully simple semihypergroups* in what follows.

Remark 20. Note that all fully simple semihypergroups have size ≥ 3, since the relation β of H is not transitive.

Example 68. An example of a fully simple semihypergroup is obtained by considering the set $H = \{a, b, c, d\}$ and the following hyperoperation:

$\circ$	a	b	c	d
a	a	$\{a, b\}$	$\{a, c\}$	$\{a, d\}$
b	a	$\{a, b\}$	$\{a, c\}$	$\{a, d\}$
c	a	b	c	d
d	a	b	c	d

The semihypergroup H has the right absorbing element a and three fully simple semihypergroups $R = \{a, b, c\}$, $S = \{a, b, d\}$, and $T = \{a, c, d\}$.

Proposition 4.3.11. *Let* $(H, \circ)$ *be a fully simple semihypergroup. If* P *is a hyperproduct of elements in* H*, then we have*

(1) $|\langle P\rangle| \leq 2$;

(2) $|P| \leq 2$;

(3) *If* $|P| = 2$, *then* P *is a simple subsemihypergroup of* H.

Proof. Since P is a hyperproduct of elements in H, $\langle P\rangle$ is hypercyclic. Obviously, $\langle P\rangle$ is simple, because H is fully simple. Therefore, by Theorem 4.3.10, the relation β in $\langle P\rangle$ is transitive. In consequence, since H is fully simple, it follows that $|\langle P\rangle| \leq 2$. The other items follow at once, because $P \subseteq \langle P\rangle$. □

Remark 21. The previous proposition can also be applied to cyclic semihypergroups generated by an element $x \in H$. In fact, since H is simple, there exist $a, b \in H$ such that $x \in a \circ x \circ b$. If we put $P = a \circ x \circ b$, then we have $\langle x\rangle \subseteq \langle P\rangle$ and $|\langle x\rangle| \leq |\langle P\rangle| \leq 2$.

4.4 THE α RELATION

Freni [19] introduced the relation γ as a generalization of the relation β. The letter γ has already been used for the corresponding fundamental relation on hyperrings by Vougiouklis [84]. Thus, there is a confusion on the symbolism. Therefore, in this section we use the symbol α instead of γ for semihypergroups. In this section, we would like the fundamental semigroup to be commutative. Notice that we use the Greek letter α for the relation because of the "A" in abelian. The main reference for this section is [19]. Therefore, we give the following definition.

Definition 4.4.1. Let H be a semihypergroup. Then, we set

$$\alpha_1 = \{(x, x) | x \in H\}$$

and for every integer $n > 1$, α_n is the relation defined as follows:

$$x\alpha_n y \Leftrightarrow \exists (z_1, \ldots, z_n) \in H^n, \exists \sigma \in \mathbb{S}_n : x \in \prod_{i=1}^{n} z_i, y \in \prod_{i=1}^{n} z_{\sigma(i)}.$$

Obviously, for $n \geq 1$, the relations α_n are symmetric, and the relation $\alpha = \bigcup_{n\geq 1} \alpha_n$ is reflexive and symmetric.

Let α^* be the transitive closure of α. If H is a hypergroup, then $\alpha = \alpha^*$ [19].

Theorem 4.4.2. *The relation* α^* *is a strongly regular relation.*

Proof. Clearly, α^* is an equivalence relation. In order to prove that it is strongly regular, we have to show first that

$$x\alpha y \Rightarrow (x \circ a)\overline{\overline{\alpha}}(y \circ a) \quad \text{and} \quad (a \circ x)\,\overline{\overline{\alpha}}(a \circ y),$$

for every $a \in H$. If $x\alpha y$, then there is $n \in \mathbb{N}$ such that $x\alpha_n y$. Hence, there exist $(z_1, \ldots, z_n) \in H^n$ and $\sigma \in \mathbb{S}_n$ such that $x \in \prod_{i=1}^n z_i$ and $y \in \prod_{i=1}^n z_{\sigma(i)}$. For every $a \in H$, set $a = z_{n+1}$ and let τ be a permutation of $\mathbb{S}_{n+1}$ such that

$$\tau(i) = \sigma(i), \quad \forall i \in \{1, 2, \ldots, n\};$$
$$\tau(n+1) = n+1.$$

For all $v \in x \circ a$ and for all $w \in y \circ a$, we have $v \in x \circ a \subseteq \prod_{i=1}^n z_i \circ a = \prod_{i=1}^{n+1} z_i$ and $w \in y \circ a \subseteq \prod_{i=1}^n z_{\sigma(i)} \circ a = \prod_{i=1}^n z_{\sigma(i)} \circ z_{n+1} = \prod_{i=1}^{n+1} z_{\tau(i)}$. So, $v\alpha_{n+1}w$ and hence $v\alpha w$. Thus, $(x \circ a)\overline{\overline{\alpha}}(y \circ a)$. In the same way, we can show that $(a \circ x)\overline{\overline{\alpha}}(a \circ y)$.

Moreover, if $x\alpha^* y$, then there exist $m \in \mathbb{N}$ and

$$(w_0 = x, w_1, \ldots, w_{m-1}, w_m = y) \in H^{m+1},$$

such that $x = w_0 \alpha w_1 \alpha \ldots \alpha w_{m-1} \alpha w_m = y$. Now, we obtain

$$x \circ a = w_0 \circ a\overline{\overline{\alpha}}w_1 \circ a\overline{\overline{\alpha}}w_2 \circ a\,\overline{\overline{\alpha}} \ldots \overline{\overline{\alpha}}w_{m-1} \circ a\overline{\overline{\alpha}}w_m \circ a = y \circ a.$$

Finally, for all $v \in x \circ a = w_0 \circ a$ and for all $w \in w_m \circ a = y \circ a$, taking $z_1 \in w_1 \circ a$, $z_2 \in w_2 \circ a, \ldots, z_{m-1} \in w_{m-1} \circ a$, we have $v\alpha z_1 \alpha z_2 \alpha \ldots \alpha z_{m-1} \alpha w$, and so $v\alpha^* w$. Therefore, $x \circ a\overline{\overline{\alpha^*}}y \circ a$. Similarly, we obtain $a \circ x\overline{\overline{\alpha^*}}a \circ y$. Hence, α^* is strongly regular. □

Corollary 4.4.3. *The quotient H/α^* is a commutative semigroup. Furthermore, if H is a hypergroup, then H/α^* is a commutative group.*

Proof. Since α^* is a strongly regular relation, the quotient H/α^* is a semigroup under the following operation:

$$\alpha^*(x_1) \otimes \alpha^*(x_2) = \alpha^*(z), \qquad \text{for all } z \in x_1 \circ x_2.$$

Moreover, if H is a hypergroup, then H/α^* is a group. Finally, if σ is the cycle of $\mathbb{S}_2$ such that $\sigma(1) = 2$, for all $z \in x_1 \circ x_2$ and $w \in x_{\sigma(1)} \circ x_{\sigma(2)}$, we have $z\alpha_2 w$, so $z\alpha^* w$ and $\alpha^*(x_1) \otimes \alpha^*(x_2) = \alpha^*(z) = \alpha^*(x_2) \otimes \alpha^*(x_1)$. □

Theorem 4.4.4. *The relation α^* is the smallest strongly regular relation on a semihypergroup H such that the quotient H/α^* is a commutative semigroup.*

Proof. Suppose that ρ is a strongly regular relation such that H/ρ is a commutative semigroup and $\varphi : H \to H/\rho$ is the canonical projection. Then, φ is a good homomorphism. Moreover, if $x\alpha_n y$, then there exist

$(z_1, \ldots, z_n) \in H^n$ and $\sigma \in \mathbb{S}_n$ such that $x \in \prod_{i=1}^n z_i$ and $y \in \prod_{i=1}^n z_{\sigma(i)}$, whence $\varphi(x) = \varphi(z_1) \otimes \cdots \otimes \varphi(z_n)$ and $\varphi(y) = \varphi(z_{\sigma(1)}) \otimes \cdots \otimes \varphi(z_{\sigma(n)})$. By the commutativity of H/ρ, it follows that $\varphi(x) = \varphi(y)$ and $x\rho y$. Thus, $x\alpha_n y$ implies $x\rho y$, and obviously, $x\alpha y$ implies that $x\rho y$.

Finally, if $x\alpha^* y$, then there exist $m \in \mathbb{N}$ and

$$(w_0 = x, w_1, \ldots, w_{m-1}, w_m = y) \in H^{m+1}$$

such that $x = w_0\alpha w_1\alpha \ldots \alpha w_{m-1}\alpha w_m = y$. Therefore,

$$x = w_0\rho w_1\rho \ldots \rho w_{m-1}\rho w_m = y,$$

and transitivity of ρ implies that $x\rho y$. Therefore, $\alpha^* \subseteq \rho$. □

Now, we want to determine some necessary and sufficient conditions so that the relation α is transitive.

Definition 4.4.5. Let M be a nonempty subset of H. We say that M is a α-part of H if for any nonzero natural number n, for all $(z_1, \ldots, z_n) \in H^n$ and for all $\sigma \in \mathbb{S}_n$, we have

$$M \cap \prod_{i=1}^n z_i \neq \emptyset \Rightarrow \prod_{i=1}^n z_{\sigma(i)} \subseteq M.$$

Lemma 4.4.6. *Let M be a nonempty subset of H. Then, the following conditions are equivalent:*

(1) *M is an α-part of H;*

(2) *$x \in M, x\alpha y \Rightarrow y \in M$;*

(3) *$x \in M, x\alpha^* y \Rightarrow y \in M$.*

Proof. (1 ⇒ 2): If $(x, y) \in H^2$ is a pair such that $x \in M$ and $x\alpha y$, then there exist $n \in \mathbb{N}$, $(z_1, \ldots, z_n) \in H^n$ and $\sigma \in \mathbb{S}_n$ such that $x \in M \cap \prod_{i=1}^n z_i$ and $y \in \prod_{i=1}^n z_{\sigma(i)}$. Since M is an α-part of H, we have $\prod_{i=1}^n z_{\sigma(i)} \subseteq M$ and $y \in M$.

(2 ⇒ 3): Assume that $(x, y) \in H^2$ such that $x \in M$ and $x\alpha^* y$. Obviously, there exist $m \in \mathbb{N}$ and $(w_0 = x, w_1, \ldots, w_{m-1}, w_m = y) \in H^{m+1}$ such that $x = w_0\alpha w_1\alpha \ldots \alpha w_{m-1}\alpha w_m = y$. Since $x \in M$, applying (2) m times, we obtain $y \in M$.

(3 ⇒ 1): Suppose that $M \cap \prod_{i=1}^n x_i \neq \emptyset$ and $x \in M \cap \prod_{i=1}^n x_i$. For every $\sigma \in \mathbb{S}_n$ and for every $y \in \prod_{i=1}^n x_{\sigma(i)}$, we have $x\alpha y$. Thus, $x \in M$ and $x\alpha^* y$. Finally, by (3), we obtain $y \in M$, whence $\prod_{i=1}^n x_{\sigma(i)} \subseteq M$. □

Before proving the next theorem, we introduce the following notations: Let H be a semihypergroup. For all $x \in H$, we set

- $T_n(x) = \left\{(x_1, \ldots, x_n) \in H^n | x \in \prod_{i=1}^{n} x_i\right\}$;
- $P_n(x) = \left\{\prod_{i=1}^{n} x_{\sigma(i)} | \sigma \in \mathbb{S}_n, (x_1, \ldots, x_n) \in T_n(x)\right\}$;
- $P_\sigma(x) = \bigcup_{n \geq 1} P_n(x)$.

From the preceding notations and definitions, it follows at once the following.

Lemma 4.4.7. *For every $x \in H$, $P_\sigma(x) = \{y \in H | x\alpha y\}$.*

Proof. For all $x, y \in H$, we have

$$\begin{aligned} x\alpha y &\Leftrightarrow \exists n \in \mathbb{N}, \exists (x_1, \ldots, x_n) \in H^n, \exists \sigma \in \mathbb{S}_n : x \in \prod_{i=1}^{n} x_i, y \in \prod_{i=1}^{n} x_{\sigma(i)} \\ &\Leftrightarrow \exists n \in \mathbb{N} : y \in P_n(x) \\ &\Leftrightarrow y \in P_\sigma(x). \end{aligned}$$

□

Theorem 4.4.8. *Let H be a semihypergroup. Then, the following conditions are equivalent:*

(1) *α is transitive;*

(2) *$\alpha^*(x) = P_\sigma(x)$, for all $x \in H$;*

(3) *$P_\sigma(x)$ is an α-part of H, for all $x \in H$.*

Proof. ($1 \Rightarrow 2$): By Lemma 4.4.7, for all $x, y \in H$, we have

$$y \in \alpha^*(x) \Leftrightarrow x\alpha^* y \Leftrightarrow x\alpha y \Leftrightarrow y \in P_\sigma(x).$$

($2 \Rightarrow 3$): By Lemma 4.4.6, if M is a nonempty subset of H, then M is an α-part of H if and only if it is union of equivalence classes modulo α^*. In particular, every equivalence class modulo α^* is an α-part of H.

($3 \Rightarrow 1$): If $x\alpha y$ and $y\alpha z$, then there exist $m, n \in \mathbb{N}$, $(x_1, \ldots, x_n) \in T_n(x)$, $(y_1, \ldots, y_m) \in T_m(y)$, $\sigma \in \mathbb{S}_n$ and $\tau \in \mathbb{S}_m$ such that $y \in \prod_{i=1}^{n} x_{\sigma(i)}$ and $z \in \prod_{i=1}^{m} y_{\tau(i)}$. Since $P_\sigma(x)$ is an α-part of H, we have

$$\begin{aligned} x \in \prod_{i=1}^{n} x_i \cap P_\sigma(x) &\Rightarrow \prod_{i=1}^{n} x_{\sigma(i)} \subseteq P_\sigma(x) \Rightarrow y \in \prod_{i=1}^{m} y_i \cap P_\sigma(x) \\ &\Rightarrow \prod_{i=1}^{m} y_{\tau(i)} \subseteq P_\sigma(x) \Rightarrow z \in P_\sigma(x) \\ &\Rightarrow \exists k \in \mathbb{N} : z \in P_k(x) \Rightarrow z\alpha x. \end{aligned}$$

Therefore, α is transitive. □

CHAPTER 5
Conclusion

The concept of a semigroup is one of the most fundamental in modern mathematics. A semigroup is a set, together with an associative binary operation. The formal study of semigroups began in the early 20th century. Semigroups are important in many areas of mathematics, for example, coding and language theory, automata theory, combinatorics, and mathematical analysis. It should be especially noted that in recent years, a sharply increasing number of articles have been published on generalizations of semigroups. Also remarkable is the increase of interest in semigroups provided with additional structures, ie, ordered semigroups.

Algebraic hyperstructures represent a natural extension of classical algebraic structures. In contemporary algebraic hyperstructures, one studies a series of algebraic objects, defined with the aid of one or several hyperoperations in a set of elements of this or that kind. One obtains one or another algebraic hyperstructure theory, depending upon the collection of these hyperoperations, their properties, and the nature of the set. One of these theories is the theory of semihypergroups. The definition of semihypergroups in general goes back at least to the 1934, one by Marty. Indeed, semihypergroups are the simplest algebraic hyperstructures that possess the properties of closure and associativity. In a semigroup, the composition of two elements is an element, while in a semihypergroup, the composition of two elements is a nonempty set. Semihypergroups have many applications in automata, probability, geometry, lattices, binary relations, graphs, hypergraphs, and other branches of science such as biology, chemistry, and physics. Many authors studied different aspects of semihypergroups, for instance, Anvariyeh, Asokkumar, Bonansinga, Changphas, Chattopadhyay, Corsini, Davvaz, De Salvo, Dehkordi, Dine, Fasino, Freni, Gutan, Hasankhani, Heidari, Hila, Jafarabadi Jafarpour, Jantosciak, Kazanci, Kemprasit, Koskas, Kudryavtseva, Leoreanu-Fotea, Lertwichitsilp, Naka, Mazorchuk, Migliorato, Mirvakili, Molaei, Mousavi, Onipchuk, Pibaljommee, Sarmin, Savettaseranee, Shabir, Spartalis, Vougiouklis, Yamak, and many others. The purpose of this book is to

Semihypergroup Theory
http://dx.doi.org/10.1016/B978-0-12-809815-8.00005-X

introduce the basic concepts of algebraic semihypergroups, together with the additional ideas that are to be developed in this monograph. The author has tried to provide a useful survey of a rapidly developing topic and the book is suitable for specialists and as an introduction to the subject for nonspecialists and graduate students.

One of the aims of this book is to extend our understanding of semigroups and semihypergroups. Every semigroup is a semihypergroup in an obvious way; however, there are many natural examples of semihypergroups that are not semigroups. The variety of subjects and problems in semihypergroup theory, and the newness of the situations in this theory are often extremely diversified and complex with respect to the classical semigroups. Semigroup theory and semihypergroup theory have developed in somewhat different directions in the past several decades. In order to study semihypergroup theory, it is necessary to know about the main concepts of semigroup theory. So, in the first chapter, we had a brief excursion into semigroup theory and presented the most important basic algebraic notions on semigroups. In the second chapter, we first explained what is meant by a semihypergroup and then gave several examples of familiar semihypergroups and discussed some of their properties. These examples showed that different semihypergroups may have several common properties. This observation provided a motivation for the study of abstract semihypergroups. Then, some basic results concerning semihypergroups were presented. We investigated regular semihypergroups, subsemihypergroups, hyperideals, quasihyperideals, prime and semiprime hyperideals, homomorphisms, regular and strongly regular relations, and simple and cyclic semihypergroups. We made an attempt to establish basic properties of semihypergroups in analogy with those of semigroups. We gave many examples and counterexamples illustrating the scope of the theory we were developing. We saw that in many respects, semigroups and semihypergroups are similar, but in many other respects they are very different. In the third chapter, algebraic properties of ordered semihypergroups were studied. We discussed several properties of ordered semihypergroups. In particular, by using the concepts of regular and strongly regular relations on an ordered semihypergroup, we answered the following question: If $(H, \circ, \leq)$ is an ordered semihypergroup and ρ is a (strongly) regular relation on H, then is the set H/ρ an ordered (semihypergroup) semigroup? In the fourth chapter, we studied the notion of fundamental relations. The fundamental relations are one of the most important and interesting concepts in algebraic hyperstructures that ordinary algebraic structures are derived from algebraic hyperstructures by them. The fundamental relation β^* is the transitive closure of the relation β studied on semihypergroups. Also, we studied the commutative fundamental relation α^*, which is the transitive closure of the relation α. Moreover, we investigated the properties of fundamental relations on semihypergroups. In particular, we presented the transitivity condition of the relation β in a semihypergroup.

The book covered most of the mathematical ideas and techniques required in the study of semihypergroups. As the first in its genre, it included a number of topics, most of which reflect the author's past research and thus provided a starting point for future research directions. Moreover, this book is the first book on this theory.

BIBLIOGRAPHY

[1] J.A. Green, On the structure of semigroups, Ann. Math. 54 (2) (1951) 163–172.
[2] S. Lajos, On characterization of regular semigroups, Proc. Jpn. Acad. 44 (1968) 325–326.
[3] S. Lajos, Notes on regular semigroups, Proc. Jpn. Acad. 46 (1970) 253–254.
[4] D.M. Lee, S.K. Lee, On intra-regular ordered semigroups, Kangweon-Kyungki Math. J. 14 (2006) 95–100.
[5] G. Birkhoff, Lattice Theory, vol. 25, American Mathematical Society Colloquium Publications, American Mathematical Society, Providence, RI, 1984.
[6] A.H. Clifford, G.B. Preston, The Algebraic Theory of Semigroups, vol. 1, American Mathematical Society, Providence, RI, 1961.
[7] A.H. Clifford, G.B. Preston, The Algebraic Theory of Semigroups, vol. 2, American Mathematical Society, Providence, RI, 1967.
[8] T. Harju, Lecture Notes on Semigroups, University of Turku, Finland, 1996.
[9] P.M. Higgins, Techniques of Semigroup Theory, Oxford Science Publications, The Clarendon Press, Oxford University Press, New York, 1992.
[10] C.D. Hollings, Some first tantalizing steps into semigroup theory, Math. Mag. 80 (5) (2007) 331–344.
[11] J.M. Howie, Why study semigroups? Math. Chron. 16 (1987) 1–14.
[12] S. Lajos, Notes on (m, n)-ideals. II, Proc. Jpn. Acad. 40 (1964) 631–632.
[13] E.S. Ljapin, Semigroups, vol. 3, American Mathematical Society, Providence, RI, 1963.
[14] J.P. Tremblay, R. Manohar, Discrete Mathematical Structures with Applications to Computer Science, in: McGraw-Hill Computer Science Series, McGraw-Hill Book Co., New York-Auckland-Dusseldorf, 1975.
[15] R.J. Warne, I-bisimple semigroups, Trans. Am. Math. Soc. 130 (1968) 367–386.
[16] O. Steinfeld, On the ideals quotients and prime ideals, Acta Math. Acad. Sci. Hung. 4 (1953) 289–298.
[17] S. Spartalis, On H_v-semigroups, Ital. J. Pure Appl. Math. 11 (2002) 165–174.
[18] O. Steinfeld, Quasi-Ideals in Rings and Semigroups, Akademiai Kiado, Budapest, 1978.
[19] D. Freni, A new characterization of the derived hypergroup via strongly regular equivalences, Commun. Algebra 30 (2002) 3977–3989.
[20] B. Davvaz, Approximations in n-ary algebraic systems, Soft Comput. 12 (2008) 409–418.
[21] B. Davvaz, Approximations in a semigroup by using a neighborhood system, Int. J. Comput. Math. 88 (4) (2011) 709–713.
[22] N. Kuroki, Rough ideals in semigroups, Inform. Sci. 100 (1997) 139–163.
[23] B. Davvaz, Polygroup Theory and Related Systems, World Scientific Publishing Co. Pte. Ltd., Hackensack, NJ, 2013.
[24] N.G. Alimov, On ordered semigroups, Izvestiya Akad. Nauk SSSR 14 (1950) 569–576.
[25] A.H. Clifford, Totally ordered commutative semigroups, Bull. Am. Math. Soc. 64 (1958) 305–316.
[26] N. Kehayopulu, On regular duo ordered semigroups, Math. Japon. 37 (3) (1992) 535–540.
[27] N. Kehayopulu, On completely regular ordered semigroups, Sci. Math. 1 (1) (1998) 27–32.

[28] N. Khayopulu, On intra-regular ordered semigroups, Semigroup Forum 46 (1993) 271–278.
[29] N. Kehayopulu, M. Tsingelis, A note on ordered groupoids-semigroups, Sci. Math. 3 (2) (2000) 251–255.
[30] N. Kehayopulu, M. Tsingelis, On subdirectly irreducible ordered semigroups, Semigroup Forum 50 (2) (1995) 161–177.
[31] N. Kehayopulu, M. Tsingelis, On weakly commutative ordered semigroups, Semigroup Forum 56 (1) (1998) 32–35.
[32] N. Kehayopulu, G. Lepouras, M. Tsingelis, On right regular and right duo ordered semigroups, Math. Japon. 46 (1997) 311–315.
[33] T. Saito, Ordered idempotent semigroups, J. Math. Soc. Jpn. 14 (1962) 150–169.
[34] T. Saito, Regular elements in an ordered semigroup, Pac. J. Math. 13 (1963) 263–295.
[35] A.H. Cliford, Ordered commutative semigroups of the second kind, Proc. Am. Math. Soc. 9 (1958) 682–687.
[36] Y.V. Hion, Ordered semigroups, Izvestiya Akad. Nauk SSSR 21 (1957) 209–222.
[37] F. Marty, Sur une généralization de la notion de groupe, in: 8th Congress Math. Scandenaves, Stockholm, 1934, pp. 45–49.
[38] W. Prenowitz, Projective geometries as multigroups, Am. J. Math. 65 (1943) 235–256.
[39] W. Prenowitz, J. Jantosciak, Join Geometries, Springer-Verlag, UTM, New York, 1979.
[40] S.M. Anvariyeh, S. Mirvakili, O. Kazanci, B. Davvaz, Algebraic hyperstructures of soft sets associated to semihypergroups, Southeast Asian Bull. Math. 35 (6) (2011) 911–925.
[41] P. Bonansinga, P. Corsini, On semihypergroup and hypergroup homomorphisms, Boll. Un. Mat. Ital. B 1 (2) (1982) 717–727 (Italian).
[42] P. Corsini, Sur les semi-hypergroupes complets et les groupoides [Complete semi-hypergroups and groupoids], Atti Soc. Peloritana Sci. Fis. Mat. Natur. 26 (4) (1980) 391–398 (French).
[43] P. Corsini, Sur les semi-hypergroupes [On semihypergroups], Atti Soc. Peloritana Sci. Fis. Mat. Natur. 26 (4) (1980) 363–372 (French).
[44] P. Corsini, M. Shabir, T. Mahmood, Semisimple semihypergroups in terms of hyperideals and fuzzy hyperideals, Iran. J. Fuzzy Syst. 8 (1) (2011) 95–111.
[45] B. Davvaz, Characterizations of sub-semihypergroups by various triangular norms, Czechoslov. Math. J. 55 (130) (2005) 923–932.
[46] B. Davvaz, Partial abelian H_v-monoids, Bull. Greek Math. Soc. 45 (2002) 53–62.
[47] B. Davvaz, Some results on congruences on semihypergroups, Bull. Malays. Math. Sci. Soc. 23 (1) (2000) 53–58.
[48] B. Davvaz, V. Leoreanu-Fotea, Binary relations for ternary semihypergroups, Commun. Algebra 38 (10) (2010) 3621–3636.
[49] B. Davvaz, N.S. Poursalavati, Semihypergroups and S-hypersystems, Pure Math. Appl. 11 (1) (2000) 43–49.
[50] M. De Salvo, Partial semi-hypergroups, Riv. Mat. Pura Appl. 17 (1995) 39–54.
[51] M. De Salvo, On the partial semi-hypergroups with empty diagonal, Acta Univ. Carolin. Math. Phys. 40 (2) (1999) 3–19.
[52] M. De Salvo, D. Freni, Cyclic semihypergroups and hypergroups, Atti Sem. Mat. Fis. Univ. Modena 30 (1) (1981) 44–59 (Italian).
[53] M. De Salvo, G. Lo Faro, Wrapping graphs and partial semi-hypergroups, J. Inform. Optim. Sci. 18 (1) (1997) 157–166.
[54] M. De Salvo, D. Freni, G. Lo Faro, Fully simple semihypergroups, J. Algebra 399 (2014) 358–377.
[55] D. Fasino, D. Freni, Minimal order semihypergroups of type U on the right, Mediterr. J. Math. 5 (3) (2008) 295–314.

[56] D. Fasino, D. Freni, Existence of proper semihypergroups of type U on the right, Discrete Math. 307 (22) (2007) 2826–2836.
[57] D. Freni, Minimal order semihypergroups of type U on the right, II, J. Algebra 340 (1) (2011) 77–89.
[58] M. Gutan, On the transitivity of the relation β in semihypergroups, Rend. Circ. Mat. Palermo 45 (2) (1996) 189–200.
[59] K. Hila, B. Davvaz, D. Jani, Study on the structure of Γ-semihypergroups, Commun. Algebra 40 (8) (2012) 2932–2948.
[60] K. Hila, J. Dine, On hyperideals in left almost semihypergroups, ISRN Algebra 2011 (2011) 8 pp., article ID 953124.
[61] K. Hila, B. Davvaz, K. Naka, On quasi-hyperideals in semihypergroups, Commun. Algebra 39 (11) (2011) 4183–4194.
[62] G. Kudryavtseva, V. Mazorchuk, On multisemigroups, Port. Math. 72 (1) (2015) 47–80.
[63] V. Leoreanu, About the simplifiable cyclic semihypergroups, Ital. J. Pure Appl. Math. 7 (2000) 69–76.
[64] S.S. Mousavi, M. Jafarpour, On free and weak free (semi)hypergroups, Algebra Colloq. 18 (1) (2011) 873–880 (special issue).
[65] S.S. Mousavi, V. Leoreanu-Fotea, M. Jafarpour, H. Babaei, Equivalence relations in semihypergroups and the corresponding quotient structures, Eur. J. Comb. 33 (4) (2012) 463–473.
[66] S.S. Mousavi, V. Leoreanu-Fotea, M. Jafarpour, $\mathcal{R}$-parts in (semi)hypergroups. Ann. Mat. Pura Appl. 190 (4) (2011) 667–680.
[67] S.V. Onipchuk, Regular semihypergroups, Mat. Sb. 183 (6) (1992) 43–54 (Russian) translation in Russian Acad. Sci. Sb. Math. 76 (1993) (1), 155-164.
[68] K. Savettaseranee, P. Lertwichitsilp, Y. Kemprasit, Regular semihypergroups of linear transformations, Ital. J. Pure Appl. Math. 26 (2009) 153–158.
[69] P. Corsini, Prolegomena of Hypergroup Theory, second ed., Aviani Editore, Tricesimo, Italy, 1993.
[70] P. Corsini, D. Freni, On the heart of hypergroups, Matematica Montisnigri 2 (1993) 21–27.
[71] P. Corsini, V. Leoreanu, Applications of hyperstructure theory, in: Advances in Mathematics, Kluwer Academic Publisher, Boston, 2003.
[72] B. Davvaz, V. Leoreanu-Fotea, Hyperring Theory and Applications, International Academic Press, USA, 2007.
[73] S.M. Anvariyeh, S. Mirvakili, B. Davvaz, Pawlak's approximations in Γ-semihypergroups, Comput. Math. Appl. 60 (1) (2010) 45–53.
[74] S.M. Anvariyeh, S. Mirvakili, B. Davvaz, On Γ-hyperideals in Γ-semihypergroups, Carpathian J. Math. 26 (1) (2010) 11–23.
[75] M. Aslam, S. Abdullah, B. Davvaz, N. Yaqoob, Rough M-hypersystems and fuzzy M-hypersystems in Γ-semihypergroups, Neural Comput. Appl. 21 (2012) 281–287.
[76] S. Chattopadhyay, A semihypergroup associated with a Γ-semigroup, An. Stiint. Univ. Al. I. Cuza Iasi. Mat. (N.S.) 56 (1) (2010) 209–225.
[77] D. Heidari, B. Davvaz, Γ-hypergroups and Γ-semihypergroups associated to binary relations, Iran. J. Sci. Technol. Trans. A Sci. 35 (2) (2011) 69–80.
[78] D. Heidari, S.O. Dehkordi, B. Davvaz, Γ-semihypergroups and their properties, Politehn. Univ. Bucharest Sci. Bull. Ser. A Appl. Math. Phys. 72 (1) (2010) 195–208.
[79] M. Jafarpour, S.S. Mousavi, V. Leoreanu-Fotea, A class of semihypergroups connected to preordered weak Γ-semigroups, Comput. Math. Appl. 62 (8) (2011) 2944–2949.
[80] S. Mirvakili, S.M. Anvariyeh, B. Davvaz, Γ-semihypergroups and regular relations, J. Math. 2013 (2013) 7 pp., article ID 915250.

[81] N. Yaqoob, M. Aslam, B. Davvaz, A. Borumand Saeid, On rough (m, n) bi-Γ-hyperideals in Γ-semihypergroups, UPB Sci. Bull. Ser. A 75 (1) (2013) 119–128.

[82] M. Ghadiri, B. Davvaz, R. Nekouian, H_v-semigroup structure on F_2-offspring of a gene pool, Int. J. Biomath. 5 (4) (2012) 1250011, 13 pp.

[83] S. Spartalis, Homomorphisms on S-H_v-semigroups, in: Constantin Carathéodory in his ... origins (Vissa-Orestiada, 2000), Hadronic Press, Palm Harbor, FL, 2001, pp. 173–178.

[84] T. Vougiouklis, Hyperstructures and Their Representations, vol. 115, Hadronic Press, Inc, Palm Harber, USA, 1994.

[85] S. Chaopraknoi, N. Triphop, Regularity of semihypergroups of infinite matrices, Thai J. Math. 4 (3) (2006) 7–11.

[86] A. Asokkumar, M. Velrajan, Regularity of semihypergroups induced by subsets of semigroups, J. Discret. Math. Sci. Cryptogr. 16 (1) (2013) 87–94.

[87] H.M. Jafarabadi, N.H. Sarmin, M.R. Molaei, Completely simple and regular semihypergroups, Bull. Malays. Math. Sci. Soc. 35 (2) (2012) 335–343.

[88] A. Hasankhani, Ideals in a semihypergroup and Green's relations, Ratio Math. 13 (1999) 29–36, hyperstructures and their applications in cryptography, geometry and uncertainty treatment (Pescara, 1995).

[89] K. Hila, K. Naka, On pure hyperradical in semihypergroups, Int. J. Math. Math. Sci. (2012) 7 pp., article ID 876919.

[90] S. Lekkoksung, On weakly semi-prime hyperideals in semihypergroups, Int. J. Algebra 6 (13-16) (2012) 613–616.

[91] S. Lekkoksung, On left, right weakly prime hyperideals on semihypergroups, Int. J. Contemp. Math. Sci. 7 (21-24) (2012) 1193–1197.

[92] J. Jantosciak, Homomorphisms, equivalences and reductions in hypergroups, Riv. Mat. Pura Appl. 9 (1991) 23–47.

[93] T. Vougiouklis, Cyclicity in a special class of hypergroups, Acta Univ. Carolin. Math. Phys. 22 (1) (1981) 3–6.

[94] J. Chvalina, Commutative hypergroups in the sense of Marty and ordered sets, in: General Algebra and Ordered Sets, HorniLipova, 1994, pp. 19–30.

[95] D. Heidari, B. Davvaz, On ordered hyperstructures, Politehn. Univ. Bucharest Sci. Bull. Ser. A Appl. Math. Phys. 73 (2) (2011) 85–96.

[96] T. Changphas, B. Davvaz, Properties of hyperideals in ordered semihypergroups, Ital. J. Pure Appl. Math. 33 (2014) 425–432.

[97] B. Pibaljommee, B. Davvaz, On fuzzy bi-hyperideals in ordered semihypergroups, J. Intell. Fuzzy Syst. 28 (2015) 2141–2148.

[98] M. Petrich, Prime ideals of the cartesian product of two semigroups, Czech. Math. J. 12 (1) (1962) 150–152.

[99] F.E. Masat, A generalization of right simple semigroups, Fundam. Math. 101 (2) (1978) 159–170.

[100] B. Davvaz, P. Corsini, T. Changphas, Relationship between ordered semihypergroups and ordered semigroups by using pseuoorders, Eur. J. Comb. 44 (2015) 208–217.

[101] Z. Gu, X. Tang, Ordered regular equivalence relations on ordered semihypergroups, J. Algebra 450 (2016) 384–397.

[102] M. Koskas, Groupoides, demi-hypergroupes et hypergroupes, J. Math. Pure Appl. 49 (9) (1970) 155–192.

[103] D. Freni, A note on the core of a hypergroup and the transitive closure β^* of β, Riv. Mat. Pura Appl. 8 (1991) 153–156.

[104] Y. Sureau, Contribution a la theorie des hypergroupes operant transitivement sur un ensemble, These de Doctorate d'Etat, 1980.

[105] R. Migliorato, n-complete semihypergroups and hypergroups, Ann. Sci. Univ. Clermont-Ferrand II Math. 23 (1986) 99–123 (Italian).

INDEX